PHILOSOPHY, CO. ...ND INFORMATIO ...IENCE

HISTORY AND PHILOSOPHY OF TECHNOSCIENCE

Series Editor: Alfred Nordmann

TITLES IN THIS SERIES

FORTHCOMING TITLES

PHILOSOPHY, COMPUTING AND INFORMATION SCIENCE

EDITED BY

Ruth Hagengruber and Uwe V. Riss

Routledge
Taylor & Francis Group

LONDON AND NEW YORK

First published 2014 by Pickering & Chatto (Publishers) Limited

2 Park Square, Milton Park, Abingdon, Oxon OX14 4RN
711 Third Avenue, New York, NY 10017, USA

Routledge is an imprint of the Taylor & Francis Group, an informa business

First issued in paperback 2016

BRITISH LIBRARY CATALOGUING IN PUBLICATION DATA

Philosophy, computing and information science. – (History and philosophy of technoscience)
1. Ontologies (Information retrieval) 2. Knowledge representation (Information theory) 3. Knowledge, Theory of.
I. Series II. Hagengruber, Ruth editor. III. Riss, Uwe, editor.
006.3-dc23

ISBN-13: 978-1-8489-3508-2 (hbk)
ISBN-13: 978-1-138-71076-4 (pbk)

Typeset by Pickering & Chatto (Publishers) Limited

CONTENTS

LIST OF CONTRIBUTORS

Selmer Bringsjord is Professor of Computer Science and Cognitive Science. He holds the chair of the Department of Cognitive Science at Rensselaer Polytechnic Institute (RPI), USA. In 1987 he obtained his PhD in philosophy from Brown University. He specialized in the logico-mathematical and philosophical foundations of artificial intelligence (AI) and cognitive science. In addition he examined the collaborative building of AI systems on the basis of computational logic. From his work on AI results the claim that the human mind will forever be superior to machines. As a Full Professor at Lally School of Management he teaches AI, formal logic, human and machine reasoning, philosophy of AI and other topics related to formal logic, as well as the intellectual history of New York City and the Hudson Valley.

Micah H. Clark is a research scientist at the Florida Institute for Human and Machine Cognition (IHMC). His research focuses on the areas of AI, computational theory of mind, socio-cognitive models of trust, psychology of reasoning, intelligence analysis and persuasive argumentation. Before joining IHMC in 2011, Micah spent fifteen years at the California Institute of Technology, NASA Jet Propulsion Laboratory, where he developed AI, system autonomy, simulation and fault management technologies for robotic interplanetary exploration systems, including the well-known Martian rovers *Curiosity*, *Spirit* and *Opportunity*. Micah obtained his PhD in cognitive science in 2010 and a joint BS in computer science and philosophy in 1999, all from Rensselaer Polytechnic Institute.

Gordana Dogig-Crnkovic is Professor of Computer Science at the School of Innovation, Design and Engineering, Mälardalen University, Sweden. She has a PhD in physics as well as in computer science. Her academic interests include the different aspects of computing as part of information processing: computing and philosophy, theory of computation and information theory. Her particular efforts in computational paradigms, knowledge generation and aspects of science of information, intelligence and cognition are remarkable. Additionally, her academic interests address the methodology and philosophy of information processing, philosophy of science, info-computationalism, ethics of computing and many more.

Luciano Floridi is Professor of Philosophy and Ethics of Information at the University of Oxford, Senior Research Fellow at the Oxford Internet Institute, and Fellow of St Cross College, Oxford. He was awarded a Cátedras de Excelencia by the University Carlos III of Madrid, was UNESCO Chair in Information and Computer Ethics, was appointed the Gauss Professor by the Academy of Sciences in Göttingen, and is recipient of the APA's Barwise Prize, the IACAP's Covey Award, and the INSEIT's Weizenbaum Award. He is an AISB and BCS Fellow, Editor in Chief of Philosophy & Technology and of the Synthese Library, and was Chairman of EU Commission's 'Onlife' research group. His most recent books are: *The Fourth Revolution: How the Infosphere is Reshaping Human Reality* (2014); *The Ethics of Information* (2013); *The Philosophy of Information* (2011); *Information: A Very Short Introduction* (2010), all published by Oxford University Press.

Ruth Hagengruber is Director of the Institute of Philosophy and holds a Chair in Philosophy at the University of Paderborn. She is also Director of the History of Women Philosophers and Scientists and Philosophy and Computing Science research areas. She studied philosophy and history of natural sciences, with a focus on the history of geometry, at the Ludwig Maximilian University in Munich. She earned her PhD in 1994 with work on the Renaissance metaphysics and mathematics in the philosophy of Tommaso Campanella. In 2011 she was awarded lifetime membership of the International Association of Computing and Philosophy (I-ACAP). Since 2012, she has been a member of Advisory Board of the Munich Center for Technology in Society at the Technical University, Munich.

Ludger Jansen teaches philosophy at the universities in Münster and Rostock, and previously held an interim professorship at the RWTH Aachen University, Germany. He received his PhD in philosophy at the University of Münster in 2001 and habilitated at the University of Rostock in 2011 with a treatise on analytic social ontology. His research focus is on social ontology and social ethics, biomedical ontology, philosophy of life sciences, philosophy of social sciences and ancient and medieval philosophy, specializing in Aristotle and Thomas Aquinas.

Jens Kohne received his MA at the Technical University (TU) in Kaiserslautern, Germany, in 2003. From 2004 to 2009, he was Research Assistant in Philosophy at the TU in Kaiserslautern, where he also obtained his PhD in 2009. His main research interests are ontology, cognitive sciences and, especially, concepts of knowledge in contemporary philosophy. He was an educational assistant at the Bildungsstätte Ebernburg in Rhineland-Palatinate from 2012 to 2013; he has been the director of the Bildungsstätte Ebernburg since June 2013.

Klaus Mainzer holds a Chair in Philosophy and Philosophy of Science at the Technical University of Munich, Germany. He obtained his PhD in 1973 and habilitated at the University of Münster in 1979. After a Heisenberg-scholarship, Mainzer was Professor of Philosophy of Science and Vice-President at the University of Constance (1980–8). In 1989–2007, he had the chair for Philosophy of Science and was founding director of the Institute for Interdisciplinary Informatics at the University of Augsburg. Since 2008, he has been Director of the Carl von Linde-Academy at the Technical University of Munich. Since 2012 he has been Director of the Munich Centre for Technology in Society at the Technical University of Munich; he is also the founder of this centre. Mainzer is a complexity scientist and has mainly worked in mathematic modelling in natural, technical and social sciences; complex systems; paradigm of self-organization; chaos theory; robotics and artificial intelligence. Moreover, he is concerned with logic; epistemology; cognition sciences; philosophy of mathematics; computer science; and natural, technical and cultural studies.

Vincent Müller is Professor of Philosophy at the Division of Humanities and Social Sciences at the American College of Thessaloniki, Greece. He earned his PhD with work on the realism debates in contemporary philosophy ('Realism and Reference') and edited a collection of papers by Hilary Putnam on the subject at Hamburg University. Currently, he is Coordinator of the EUCog Network, the 'European Network for the Advancement of Artificial Cognitive Systems, Interaction and Robotics'. This network consists of roughly 800 European researchers in artificial cognitive systems and is supported by the European Commission. His research focuses on the future of computational systems, particularly on the prospects of artificial intelligence.

Uwe V. Riss is Senior Researcher in the New Assets and Business Model Innovation group of SAP (Switzerland) AG. He obtained a PhD degree in theoretical chemistry from Ruprecht Karls University of Heidelberg and a degree in Mathematics from Philipps University of Marburg. Currently he is Coordinator of the EU-funded Mobile Cloud Network project. In his research he focuses on business model innovation and knowledge management. In addition, he works on the philosophy of scientific and mathematical practice with a particular focus on the philosophy of Michael Polanyi.

Barry Smith is Distinguished Professor of Philosophy and Adjunct Professor of Neurology and Computer Science in the State University of New York at Buffalo, USA. He is also Adjunct Professor of Neurology and Computer Science. In 1976 he obtained his PhD from the University of Manchester with a dissertation on the ontology of reference. He is also Research Scientist in the New York State Center of Excellence in Bioinformatics and Life Sciences. From 2002 to

2006 he was Director of the Institute for Formal Ontology and Medical Information Science (IFOMIS) in Leipzig and Saarbrücken, Germany. In 2002, Smith received the Wolfgang Paul Award of the Alexander von Humboldt Foundation in recognition of his scientific achievements. In 2010, he received the first Paolo Bozzi Prize for Ontology from the University of Turin. Smith's primary research focus is ontology and its applications, especially in biomedical informatics where he has worked on a variety of projects relating to biomedical terminologies and electronic health records, and most recently in military intelligence.

Holger Andreas is an Academic Assistant at the Seminar for Philosophy, Logic and Philosophy of Science at the Ludwig Maximilian University of Munich. He obtained a BSc in physics in 1996, an MA in philosophy and physics in 2001, and a PhD in philosophy in 2005 from the University of Leipzig. One central result of his dissertation in philosophy consisted of a new formal semantic field of theoretical terms. Following a year as Visiting Scholar in Stanford, he was active as an Academic Assistant first in Bonn, and has been in this role in Munich since 2009. His research interests cover the philosophy of science, logic, epistemology, cognitive science and philosophy of mind.

Francis C. Dane is a social psychologist and received his PhD in social psychology from the University of Kansas in 1979. As Professor he currently holds the Chair of Arts and Science at the Jefferson College of Health Science. His research encompasses mathematical models of medical outcomes and environmental determinants of mastery learning.

Frederico Fonseca is currently the Co-Director of the Centre for Online Innovation in Learning at Penn State University. He obtained his PhD on spatial information science and engineering at the University of Maine in 2001. At Penn State University he was the Associate Dean for Education of the College of Information Sciences and Technology. His work deals with the flow of information starting from its conceptualization in human minds to its implementation in computer applications.

Klaus Fuchs-Kittowski was Professor of Information Processing at the Humboldt University in Berlin where he worked for more than twenty years. He earned a PhD in philosophy on the problem of determinism with regards to the technical regulation of living organisms. In 1994, he was among the founders of the Computer Centre of Humboldt University. In 1971, he was awarded the Rudolf Virchow Prize for medical research on determinism and cybernetics in molecular biology. He received the IFIP Silver Core and became elected member of the Leibniz-Sozietät der Wissenschaften. His major research focus has been on the theory and methodology of information system design.

Kai Holzweißig Kai Holzweißig holds a PhD in computer science and a MSc in cognitive science. Presently, he works for Daimler AG in the Information Technology Management division. Kai is an adjunct lecturer at Reutlingen University. His research interests are in information systems design, socio-informatics and interaction design.

Ludwig Jaskolla is Research Assistant at the Munich School of Philosophy, Germany. He acquired a MA in philosophy, logic and cultural studies in 2008 and is currently pursuing his PhD, working on the thesis 'Real Four-dimensionalism – An Essay on the Ontology of Time and Persistence'. His work focuses on analytic metaphysics, philosophy of mind and philosophy of religion.

Jakob Krebs is Research Assistant in Philosophy at the Goethe University Frankfurt, Germany, and received a PhD in 2012 from this institution. His research focuses on philosophy of information, metaphorology, philosophy of language, philosophy of mind, epistemology and philosophy of media. He also manages the institutional e-learning activities and other digital transformations.

Jens Krüger is Visiting Scientist at the Heinz Nixdorf Institute of the University of Paderborn. At present he works for Daimler AG.

Tillmann Pross received a university degree in linguistics in 2005 and a PhD in general and computational linguistics in 2009 from the University of Stuttgart. Then he worked on semantics and pragmatics at the Institute for Natural Language Processing (IMS). The application scenario of his dissertation – 'Grounded Discourse Representation Theory' – was the interaction between humans and robots.

Matthias Rugel is Research Assistant at the Munich School of Philosophy, pursuing his PhD following the thesis 'Panexperientialism: The Sensations of Atomic Units as Blueprint for Mind, Matter and Reality'. He concentrates his academic efforts on analytic metaphysics, philosophy of mind, software-based ontology, philosophy of religion and social philosophy.

David J. Saab is Researcher and Instructor at the Penn State University. He gained a PhD degree in information sciences and technology at Penn State University in 2011. His primary research is focused on information system ontologies as manifestations of cultural schemas. He is also interested in information metaphors and visualization and the philosophy of information.

Joshua Taylor is a Research Engineer, who has worked for Assured Information Security Inc. since 2011. Prior to this he worked at the Rensselaer Polytechnic Institute for more than eight years. At this institute he also had obtained a MS in computer science in 2007 and a PhD in 2013. Joshua's areas of focus are machine learning, artificial intelligence and semantic technologies.

Uwe Voigt is Professor and Chair for Analytic Philosophy and Philosophy of Science at the University of Augsburg, Germany. In 2007 he habilitated with his treatise on 'Aristotle and Information Terminology: An Ancient Solution for a Contemporary Issue?' in Bamberg. His major academic interest concerns the reflection on multidisciplinarity, systematically qualifying philosophical classics on their present relevance.

Aziz F. Zambak is an Assistant Professor at the Department of Philosophy at Middle East Technical University, Ankara. He received his PhD in Philosophy from Katholieke Universiteit Leuven, Belgium in 2009. His areas of specialization include artificial intelligence, philosophy of information, computational ontology and logic.

LIST OF FIGURES AND TABLES

INTRODUCTION: PHILOSOPHY'S RELEVANCE IN COMPUTING AND INFORMATION SCIENCE

Ruth Hagengruber and Uwe V. Riss

I

The relevance of computer and information science for today's life is obvious, whereas it seems to be less obvious whether this also holds for the philosophy in this field. The velocity of technological development has left no space for questions that concern the foundations of information and computation. However, a closer look reveals that computer and information science are thoroughly steeped in philosophical assumptions, even though this fact rarely stands out in public awareness. It only comes to the fore when technical developments slow down or miss our expectations. Nevertheless, the awareness is growing that it might be time to establish an exchange between the technical and philosophical disciplines. The main difficulty that we have to overcome in starting this process consists in the historical misunderstandings and mutual distrust on both sides that have often disturbed the dialogue.

While nobody seriously doubts that there are social and historical dependencies between technology and science, the philosophical impact on science and technology is often disputed or even completely denied. In fact, the genuine philosophical procedures of analysis and synthesis play an eminent role in science and technology. Definitions, rules and laws, by which scientific functionality and the realm of its applicability is determined, clearly prove philosophy's impact in this respect. The aim of this book is to clarify these connections to philosophy, showing philosophy's relevance in various disciplines, which are constitutive to information and computation sciences (IS/CS) and hence and finally to its application within information technology, exposing its relevance even to the practitioner.

As *scientific* disciplines, information and computation science have to strive for reliable foundations. This book will support the search of these young sciences to find their place among older and more established disciplines. Here the question might come up to which extent we have to take the difference between information and computation science into account. We have to ask if there is a need to clarify the relevance of this distinction concerning the attempt at analysis offered here. At this stage of research we are convinced that – regarding their philosophical foundation – the two sciences go mainly hand in hand even though their respective approach towards philosophy might be different. Future discourse might handle the philosophical foundation of both strands separately, however, for the time being it appears to be convenient to consider them together. The contributors refer to IS and/or CS, respectively and according to their particular subject, which determines their perspective towards the investigation of philosophy's relevance in their respective area. For all contributors philosophy is the common focus and unites the views of the involved disciplines.

Important questions of ethics in IS/CS are not dealt with in this book. This is not because the editors vouch for a position which does not give ethics an eminent rank. We are, rather, convinced that ethics is at the basis of all judgments and actions. Sciences and technical practices are built upon decisions which result from moral reflections. It is also true that the public is deeply aware of the ethical implications of IS/CS. This field has become a huge area of discussion.[1] However, we decided not to include ethical questions in the present volume. In keeping with this book's main purpose, it only includes contributions that focus on ethical provisions for practitioners.

II

When philosophers started doing philosophy in ancient times, they began by posing the question of what knowledge is. They then discussed how the difference of knowledge (episteme) and *techne* (tecnh) became characteristic of scientific development.[2] Many philosophers and scientists still maintain this distinction and for them philosophy and IS/CS represent different ways of knowing. Philosophers and practitioners become separated from each other, as if one could do without the other, a view that Greek philosophers such as Socrates would never have agreed to. This separation led far further.[3] This separation between different types of knowledge and doing shaped different terminologies in sciences and handcrafts, that is, in practical knowing. We have regarded it as our task to recall this starting point of European sapience of the joined endeavour of philosophy, science and *techne*. We must not forget the integrated perspective that stood at its beginning and which must be seen as the reason for the success of modern scientific and technical development. However, the separation into different sciences has also been an essential precondition for this success. Philosophy, science and *techne* are reciprocally bound to each other but built on their own respective strengths. If we want to understand the barriers for an exchange between philosophers, scientists and practitioners today, we must look back into a history of more than two thousand years.

In his famous dialogue *Meno*, Plato questions the various ways of knowledge. The discussion arises when Socrates asks if for a *successful trip to Larissa* it is necessary to *know* the way to Larissa.[4] The distinction concerns the differentiation and the dependency of bodily experience, necessities and contingencies. Similar ideas still came up within the artificial intelligence (AI) discussion some decades earlier, for example, as Hilary Putnam's thought experiment of a *brain in a vat* demonstrates.[5] Putnam states that knowledge is not bound to physical entities, holding to the conviction that knowing is a 'disembodied' transformation of data and signals. In his *Engine of Reason, the Seat of the Soul* (1995) Paul Churchland tried to demonstrate that machines perform knowledge processes, in the attempt to confirm that knowledge is not a sort of spiritual and non-bodily power but originates from algorithms and adaption strategies.[6] Before that, Simon and Newell had described the heuristics of invention.[7] Since then, enormous efforts in philosophy and artificial intelligence have been undertaken to understand the *synthesis* of mental processes and actions. The concept of the *embodied mind* has led to a multitude of developments within robotics and related fields of research that emerged from the interdisciplinary studies of robotics, cognition and philosophy.[8]

Another controversial philosophical issue in the intersection of philosophy, science and *techne* is *objectivity*, defended by philosophers over centuries and criticized by philosophers and finally abandoned by nineteenth-century positivism and pragmatism. The philosophical idea of conceptualizing a kind of knowledge expected to be independent of contingencies and subjective arbitrariness, influences the tradition of science and practice in many fields. Yet, it can even be seen as one of the main pressing forces of the idea of science.

A third influential concept is the nineteenth-century *separation of natural and technical sciences from humanities*, as it was articulated in the philosophy of Dilthey and others. Quite a number of influential philosophers in the twentieth century adopted it and even aggravated it.[9] Based on Heidegger's criticism of technology and influenced by Adorno and Horkheimer, philosophers attacked blind confidence in technology or even harshly criticized the influence of technological development in general, following Adorno's perspective by talking about the 'Disenchantment of Nature'.[10] Dessauer (1927) and the outstanding Cassirer (1930) took a more rational approach towards technology and started a discourse on the cultural consequences of technology by means of anthropological categories.[11] The ideas of the latter, in particular, are not yet intensively examined in the philosophy of information science. Others saw philosophy as one science among others. Neurath, Carnap, Reichenberg and other logical empiricists before them had even inverted the direction of philosophical research and demanded a *scientific* approach in philosophy, transforming philosophy into a branch of science.[12] Stegmüller and Quine explained that there are no specifically philosophical problems at all.[13] Feyerabend, Kuhn and others referred to the pragmatic idea of usability and criticized the claim for a specific way of knowing.[14] These and similar ideas were controversially discussed when computer science came into being, even though they were not explicitly taken into account at that point. Many scientists and practitioners in the field of IS/CS would not even consider a historical perspective and even less a philosophical one. They see the genesis of IS/CS in the discussion between Gödel and Turing. Of course there is good reason to do so, however, such perspective only gives us a fragmentary insight into the constitutive relationship of philosophy and computing.

Today's philosophy of computer and information science mainly aims at establishing a foundation of these new technological disciplines. The first approaches in this respect started with the attempt to establish the foundations of artificial intelligence. A prominent contribution that can be associated with this endeavour is Winograd's and Flores's *Understanding Computers and Cognition* (1986).[15] In this book, the authors investigated the influence of different philosophical positions on our understanding of artificial intelligence. It has led to a fertile criticism of the assumptions on which early artificial intelligence research programmes were based.[16] This discussion has definitely enhanced our understanding of intelligent behaviour and inspired new approaches which take the actual interactions of robots with their environment into account.[17] Thagard's *Computational Philosophy* (1988) again advocated for a fertile exchange between computer science and philosophy.[18] He realized the necessity of epistemological considerations in science and encouraged the reflection of scientific results in philosophy.

Another area in which we find a significant influence of philosophy in computer science is human–computer interaction. It concerns the nature of human communication and its hidden assumptions that often cause people to misunderstand the computers they work with since they expect them to react in the same way as an intelligent human communication partner. A prominent example of such investigation is Dourish's *Where the Action Is* (2001), in which he refers to Heidegger and Wittgenstein, whose positions on cognition, language and meaning led to a new understanding of the interaction of human beings and machines.[19] We also find a strong inspiration from and reference to philosophy in activity theory,[20] which is based on the works of Vygotsky and Leont'ev, whose theories are rooted in dialectical philosophy.[21] Their theoretical perspective has helped to clarify the role of information technology in specific settings and work situations that are characterized by the use of information technology as a tool.[22]

During the last decades another area, in which philosophical topics play a central role, has emerged concerning the development of information systems. Here, the question has been raised to which degree these systems can be further developed to knowledge management systems. This centrally addresses the question of knowledge representation and handling. One of the fundamental books in this respect is Nonaka's and Takeuchi's *The Knowledge Creating Company* (1995),[23] in which the authors essentially build their approach on the philosophy of Ryle and Polanyi's concepts of implicit and explicit knowledge as well as on the Japanese philosophical tradition.[24]

Relevant philosophical questions come to the fore in the context of the Semantic Web discussion and with the rise of semantic technologies in general.[25] The idea was anticipated in the early 1990s when Gruber introduced the design principles of formal ontologies.[26] The design of ontologies raised questions about the philosophical foundations of the underlying models and led to a vivid discussion on the topic. Since then, we can observe the generation of a plethora of ontologies in various domains as well as the emergence of ontological research programmes. In particular, the observations of incompatible coding mechanisms and conceptual inconsistencies have shown that a revision of the fundamental assumptions seems to be necessary. It was Barry Smith who started such investigation and used his philosophical ideas to concretize ontological projects such as the development of the Basic Formal Ontology (BFO).[27]

Finally, we can see that philosophy of information and information science is closely related to philosophy of computation. Both have to deal with the problem of the distinction between information and data. Historically speaking, information science found its roots in library science, but was then strongly influenced by the development of information technology. Its relation to computer science can be essentially traced back to Shannon's and Weaver's *Mathematical Theory of Communication,*[28] but recently gained increased interest due to the development of a philosophy of information, to which Luciano Floridi has decisively contributed over the last few decades.[29] Philosophy of information has also led to a new discussion about the role of language as one of the main tools of information transfer. The topic is not completely new and a discussion about the status of language can already be found in the works of Leibniz, who tried to define a merely philosophical (i.e. rational) language. In this work, he not only tried to constitute language as a game of rule-directed symbols, but as a reflection of the structure of reality. Such mirror theory that identifies the structure of reality and an (ideal) language has also been the aim of Wittgenstein's *Tractatus logico-philosophicus* (1922).[30] However, Wittgenstein had already realized the irredeemability of this endeavour, so that mirror theory is mainly abandoned today.[31] Nevertheless, it decisively influenced early research in artificial intelligence, and influences the discussion of the concept of information to this day.

In addition to this work-related recapitulation of the exchange between philosophy and sciences in their relation to computation and information, we can also look at the organizational side of this exchange. Meanwhile, the philosophy of computation and information had established its footing in proper associations and conferences. What originally came along as 'Computer-Assisted Instruction' at various philosophy conferences was further fostered by the American Philosophical Association (APA) through its Committee on Philosophy and Computers, and finally resulted in the foundation of the International Association for Computing and Philosophy (IACAP) in 2004. The most prominent expression of the constantly growing interest in the topic is a series of regular international conferences that started in the 1980s.

Another clear indicator of the increasing attention of philosophers to information and computation is the number of articles in the *Stanford Encyclopedia of Philosophy*. The most prominent ones are Floridi's 'Semantic Conception of Information',[32] Turner's and Eden's 'The Philosophy of Computer Science',[33] Barker-Plummer's 'Turing Machines',[34] Immerman's 'Computability and Complexity',[35] Horst's 'The Computational Theory of Mind',[36] Bynum's 'Computer and Information Ethics',[37] among many others.

If we take a closer look at the areas of philosophy that are discussed at these conferences, we find that the topic almost covers all branches of philosophy. Therefore the selection of fields which we have chosen in this collection cannot be complete. Nevertheless, we intend to cover a broad and representative spectrum of the currently discussed issues. Examples of the questions which we address in this compilation are:

- What do we mean by computation and information?
- Is the complexity of human thinking and computing the same?
- What does the term 'formal ontologies' refer to?
- What is the relation between knowledge and its formal representations?
- Is computation more than what we do with computers?
- To what extent do informational models influence our action and vice versa?
- Can human beings and computers coexist without conflicts?

There are many open questions in contemporary debates, and all of them require an extensive discussion. Philosophy offers various valid positions towards them, and the discourse which develops from an exchange of arguments definitely represents a significant progress for philosophy as well as for computation and information science. Some of these research areas are already established and describe the (historical) core of the dialogue between philosophers and computer and information scientists, while others rather address the evolving questions such as the philosophical study of complexity and action theory. These two areas are gaining increasing interest, and we will take a closer look at their recent development.

The theory of complexity or dynamical systems originates from physics, where it has been developed to explain the evolution of dynamical systems, which are represented by systems of coupled differential equations. One particular focus of interest has been the relation of complex systems and chaos theory.[38] In the course of the development of complexity theory it has become clear that the same concepts could also be applied to cognitive processes.[39] Indeed, the obvious complexity of brain processes apparently suggests such an approach. The further development in this area has led to the idea of swarm intelligence and swarm robots.[40] An extension of this idea is the concept of info-computationalism, which refers to system dynamics as a means of understanding the universe and its development.[41]

The second novel approach in the philosophy of computation and information is action theory. It addresses the philosophical question of how tools influence human action and what it means to regard a computer as well as symbolic systems as tools. One of its starting points has been the observation of the entwinement of action and knowledge as it is brought forward by the concept of practical knowledge or know-how.[42] In addition, the notion of knowledge also plays a role in social practices where the interest concerns knowledge transfer and competence of coordinated action. The insight in the connection between knowledge and action goes back to Aristotle who described the distinction of *techne* as knowledge for production and phronesis as knowledge for valued rational action. In today's philosophy, we find a continuation in the discussion between epistemic intellectualists[43] and anti-intellectualists[44] on the question of whether practical or propositional knowledge is more fundamental and whether one can be reduced to the other. This controversy has decisive consequences for computer science since it concerns the question of whether intelligent behaviour can be exclusively based on knowledge representations such as ontologies and the application of formal logic or if it is based on complex system dynamics and swarm intelligence.

In this wide field of possible topics, it seems to be a daring endeavour to address them all together, an endeavour that is actually impossible. Therefore the contributions to this essay collection concentrate on specific topics, which reflect the most important questions concerning the relation of philosophy to computation and information science. Ontology, complexity and knowledge representation can be seen as classical topics that have already engaged in some dialogue with philosophy. In contrast, action theory and info-computationalism represent some of the novel areas of research that are discussed.

The first part of the dialogue deals with the concept of computation and information. This topic is illuminated from two sides. Luciano Floridi, one of the most influential thinkers in the philosophy of information and technology, will give an insight into the development of the theory of information and its relation to the external world. Being conversant with the origins of our philosophical thinking in ancient Greece, he looks back on the decisive transitions in the development of human culture and technology. In his retrospective he identifies these decisive transitions: the Copernican revolution that moved the earth out of the centre of the universe, the Darwinian revolution that moved the human race out of the centre of the universe, and the Freudian revolution that demonstrated that our self-perception is not transparent. It is this background against which we have to make sense of the informational revolution in which we are currently involved.

While Floridi takes a macro-perspective to explain the role of information in history, Jakob Krebs adopts a micro-perspective and discusses the idea of trans-ferability of information and what it actually means for the recipient and the sender of information. If we go back in history, for example to Shannon and Weaver,[45] we find a different idea of information, which Qvortrup called the *substantial* understanding of information.[46] This substantial view regards infor-mation as a thing which is simply transported from the recipient to the sender. Krebs explains why this simplified view is incorrect. Referring to relational informativeness, he points to the importance of prior knowledge and situational context for the interpretation of data for the resulting information. We can also express the respective views of the first two chapters by saying that Floridi exam-ines to what extent information constitutes the basis of life for human beings, whereas Krebs examines how human beings reversely constitute information.

The following contribution by Uwe Voigt deals with the concept of infor-mation and with the different meanings of this concept. If there are different meanings, we have to answer the question about the relation between them. To demonstrate the variance he refers to Ott,[47] who identifies eighty more or less specific definitions of information, the most famous of which is probably Bate-son's 'difference that makes a difference'. In order to resolve the confusion about the concept of information, he compares it to the concept of life, as Aristotle has discussed it. In the same way as Aristotle was content with two concepts of life, Voigt concludes that the existence of different meanings of information might also be natural and acceptable, expressing a certain bipolarity in the con-cept, which represents at least two sides: a substance related and a process related one. This observation of bipolarity in the concept of information will reoccur in the discussion of knowledge representations, which show a static formal and a dynamic action-related side.

Klaus Fuchs-Kittowski provides the concluding contribution in this section, in which he recapitulates the tension between computer science and philosophy during the course of their coexistence. His chapter again provides an overview of all intellectual approaches that have influenced the development of computer sci-ence with its ups and downs, in particular concerning the development of artificial intelligence. However, Fuchs-Kittowski does not restrict his view to philosophy and computer science only but also looks at other scholarly areas in their periphery. He finally turns to the concept of noosphere and the influence that philosophy and computer science have had on the integration of information and communication technologies in the processes of social and individual development.

The following two essays deal with the previously mentioned topics of complexity and systems theory. They concern the question of whether the classical paradigm of computation, which is based on predefined symbolic representations and provides us with a deterministic understanding of natural processes, reflects the actual nature of processes such as the one that we observe in the human brain. Klaus Mainzer argues that our idea of deterministic computation is too restrictive in this respect. Therefore, we need more open computing paradigms that allow freely interacting computational agents. Such approaches seem to be more promising to bring artificial intelligence forward. These models reveal that free computational agents allow for the emergence of intelligent behaviour. However, such increased freedom also means a limitation of control. Human beings who work with such intelligent machines have to take over more responsibility. Similarly, Aziz Zambak deals with the question of the conditions of artificial intelligence; however, he refers to our intentions in this respect. Such intentions can be used to build application, to develop alternative forms of intelligence, to copy human intelligence, or simply to provide machines that coexist with human beings in a symbiotic manner. He stresses that agency, understood as a means of direct interaction with the concrete world, is a crucial feature of any application that is expected to behave in an intelligent way. Only the complexity of reality provides a test bed that is rich enough to train such intelligence. In the latter conclusion, both contributions come together again since they regard real-world complexity as a source of *friction* necessary to produce artificial intelligence.

The central topic of the next four contributions is formal ontology, the development of which is closely related to semantic technologies in computer science. It is interesting to note that Quine decisively influenced model building and the representation of reality. Usually, the foundational starting point of ontology construction as an important research field in information science can be seen in Gruber's definition of an ontology as 'an explicit specification of a conceptualization.'[48] This definition was mainly driven by practical needs and lacked philosophical analysis. Barry Smith refers to this open point and addresses the requirements and conceptualization of formal ontologies in his contribution. Based on his analysis, he argues for an ontological realism.[49] In order to provide a sound basis for ontologies he refers to the history of ontology as a philosophical discipline. He recalls the different schools of philosophical ontology represented by substantialists and fluxists, who debated the question of whether ontology is based on objects or processes. Even today this distinction remains relevant since process-based ontological theories are a minor, but vivid part of ontological research.[50] He identifies a second line of division between adequatists and reductionists. The latter group reduces reality to an ultimate level of entities that compose the 'rest' of the universe. In contrast, Barry Smith favours adequatism that allows transcending substantialism and fluxism. With these explanations, he points to the subtleties of philosophical ontology, of which most practical ontologists are often unaware, so that they overlook pitfalls of ontological analysis.

Jens Kohne goes back to the origins of ontology and describes the role of ontological categorization for our understanding of reality. He turns to the controversy between realism and nominalism in ontology and asks what its relevance is for today's information science. The question is whether representations in information science describe mind-independent entities or whether these representations are also influenced by the subjective perspectives. The latter aspect already appeared in the case of terminology, in which at least cultural and linguistic factors influence each object representation. Finally, he poses the question of how we actually can access reality and which are the consequences for representations in information science.

The fourth ontology-related contribution goes back to application. Ludwig Jaskolla and Matthias Rugel present the development of an ontology of questions and answers. In their approach they deal with surveys and, as is usually the case in social sciences, with population. They place particular emphasis on the objects of the survey and the people who partake in it. Philosophy comes into play by contrasting a realist and an anti-realist interpretation of populations; the anti-realist position assumes that subject and object population are not clearly separated, whereas the realist position claims the opposite. Jaskolla and Rugel argue for the realist position which appears to be more convincing to them.

As we recognize from these four contributions, the main problem is to clarify what we actually describe by ontologies: conceptualizations or linguistic phenomena versus real entities. Most of the authors tend to the realist position, in which problems such as vagueness or ambiguity seem to become irrelevant. The foremost goal is to describe what is given, independently of a particular representation. However, as the case of terminology shows, in many instances it is difficult to get rid of the influence of language, even if one attempts to do so. However, it seems that in some domains such as physiology, in which medical ontologies are developed, it is possible to describe entities as they are in a clear way. In other cases, in which the human perspective plays a more prominent role, this appears challenging, to say the least. Generally, it appears to be necessary to reflect on the particular conditions that allow ontologists to develop mind-independent representations and what this independence actually means.

The question of ontologies is closely related to that of knowledge representations. However, instead of representing reality, the aim here is to represent knowledge as a specific human capability. As in the case of ontologies, we have to deal with the question of whether the object of this representation is an objective or a subjective entity, and we have to investigate its characteristics.[51] The answer to this question is crucial for knowledge management and other application areas dealing with human knowledge. Early attempts to grasp human knowledge resulted in the classical philosophical knowledge definition of justified true belief. It shows the particular focus on propositional knowledge, which was regarded as the one specific for human beings. The definition assumes that propositional knowledge is naturally represented by language, so that it can be simply codified and stored in IT systems. However, the validity of the definition was fundamentally challenged by a class of counterexamples, the so-called Gettier cases,[52] which showed that the nature of knowledge is more complex as it is reflected in the justified true belief definition. Gettier cases yield a crucial result of modern epistemology, which also challenges the attempt to codify knowledge and store it in IT-based management systems. Most scientists who criticized the latter attempt point to the non-explicit character of most knowledge, referring to the work of Ryle and Polanyi.[53] It was particularly Polanyi who had stated that all explicit knowledge is rooted in implicit knowledge and that the latter is not necessarily accessible to codification. Despite these critical voices, there is still a prominent group of practitioners and even researchers who assume the validity of the traditional definition.

Knowledge representations are not only important for knowledge management systems, but also play a decisive role in artificial intelligence. In this area, many researchers argue for the traditional approach towards intelligence, which consists of formal knowledge representations and the application of fixed reasoning rules as the most promising way to simulate intelligence. Selmer Bringsjord, Micah Clark and Joshua Taylor present reflections on this view in their contribution on knowledge representations and reasoning. They argue for a stronger reflection on philosphy in the endeavour of applying knowledge representation and reasoning in the realms of mathematics and socio-cognition. Although they apply a formal approach to deal with this task, they are aware of the fact that we can only expect to achieve rather limited capabilities of intelligent machines in this way if compared to human minds.

The second contribution by Holger Andreas deals with frame systems, which were introduced by Minsky to represent knowledge.[54] This framework for dividing knowledge into substructures describes stereotyped situations based on Minsky's original idea to grasp meaning by exploiting Chomsky's work on syntactic structures.[55] More specifically, Holger Andreas shows the relations between frames and scientific structuralism.[56] According to this paradigm, scientific theories are model-theoretic nets that are associated with scientific concepts that represent empirical systems by set-theoretic entities. The approach follows the idea that animals, human beings and artificial systems mainly use knowledge representations as basis for the interaction with their environments. If representation is a precondition of problem solving, models must cover relevant features of the environment. In particular, he connects this view with recent technical approaches towards semantic representation such as the Resource Description Framework Schema (RDFS). Andreas's chapter demonstrates in which way this knowledge representation fits structuralist reconstruction of reality and proposes a prototypical application of his approach.

David Saab and Frederico Fonseca highlight the cultural background of knowledge representations, which is often neglected. They continue the investigation of ontologies, but concentrate on the general aspects of representation. While syntax is mainly independent of the respective context, the semantics of representations show a high variability in terms of different settings. This is one major reason why knowledge representations fail to provide valid results. To illustrate this, they refer to Heidegger's phenomenological examination of ontology and the use of his notion of being-in-the-world.[57] They argue that Heidegger's philosophy shows that the distinction between subject and object, as introduced by the Aristotelian categorical notion of ontology, becomes actually blurred if we take the concrete setting of a situation into account. In this respect the cultural background of the subjects who interpret such representations is important. To explain their view, they refer to connectionist theory and the notion of the cultural schema.[58] Such schemas play a decisive role in the comprehension of knowledge representations as well as in the understanding of the concept of information.[59] In contrast to traditional knowledge representations, the latter are not fixed and can appear in different configurations reflecting the underlying cognitive processes in a situation. Culture becomes manifest in such schemas representing Heidegger's ready-to-hand background. It is argued that in order to establish successful communication, knowledge representations must always refer to the underlying shared cultural schemas.

The following section about action theory addresses the non-representational aspects of information and knowledge. It is based on the insight that both information and knowledge are closely related to actions of communicating information and actualizing knowledge by its application in conrete situations.[60] Peter Janich has described action as an actualization of a scheme[61] and used this idea to build a bridge to representation.[62] The philosophical task is to analyse the role of action in this process. In order to describe and understand the exchange of information, it is important to know how communciation works.[63] Action theoretic approaches provide a critical view of the naturalized understanding of information. This is based on a discussion which Janich and Ropohl started some time ago and in which they explained that, although information is a key concept in today's sciences, its meaning is still unclear.[64]

In this volume, Uwe Riss points at the fundamental difference between abstract knowledge representations and concrete actions, which has been identified as one of the major barriers for knowledge management.[65] As already mentioned, the justified true belief definition of knowledge relies on the idea that knowledge is mainly propositional and can be properly represented by means of language. However, if we consider the relation between knowledge and action more carefully,[66] we find non-propositional (i.e. practical) forms of knowledge. To reflect this fact, Kern characterized knowledge as rational capacity which is actualized in action.[67] Riss examines to what extent the fundamental gap between abstract representation and concrete action can be overcome. The described difference refers to the same set of problems that Saab and Fonseca raise in their discussion of the role of a specific cultural background for the interpretation of knowledge representations. The central question concerns the transformation of abstract knowledge representations into concrete actions, which has to take the specific context and the individual capacities of acting subjects into account.[68] The nature of the gap is explained on the basis of various examples based on Wittgenstein, Ryle and Polanyi. It is shown that knowledge is not simply transported from one place to another, such as the physical manifestations of knowledge suggest, but that the transfer of knowledge requires an implicit reconstruction process, in which the hierarchical structures of knowledge reflect an analogous structure of action.[69] Analysing the analogy of both structures, we can explain the specific relevance of the individual constituents of the justified true belief conception for action. On the basis of this investigation, the consequences for the design of knowledge management systems are indicated. It is argued that one possible way to take the relation between knowledge representations and actions into account is the use of task management systems, in which concrete actions and abstract representations can be closely entwined. Riss emphasizes that the actual central idea of this approach is not the inclusion of a formal representation of action, which shows the same deficiencies as general knowledge representations, but the provision of an action-adapted environment that is involved in the execution of action and influenced by the actor's prior experience. The rationale described in the essay shows in which way philosophical analysis can inspire new approaches and designs in information technology.

Kai Holzweißig and Jens Krüger examine the relation between knowledge and action from the viewpoint of new product development. The development processes are based on well-defined process models to support the involved actions as efficiently as possible. In particular, they point at the connection of these process models and the experts' personal knowledge, which determines the success of the production process. This, however, requires the fine-tuned coordination and mutual understanding of all involved parties. Looking at the prevailing positivist paradigm in rationalizing the development process, Holzweißig and Krüger remind us of the difference of information and data,[70] which positivists tend to neglect. This point resembles Riss's argumentation, according to which we have to clearly distinguish abstract objects from concrete situations and have to carefully investigate their interplay, which they describe in their two-level model. As a concrete measure to support the production process at these two levels, they propose the use of information technology in social software, the wiki being the most popular in this respect.[71]

Tillmann Pross starts his investigation from the specific interaction of human beings and machines and discusses the role of discourses herein.[72] He examines the logical form of action sentences with respect to individuals acting under the condition of time. This condition is reflected in the temporal profile of phrases that go beyond simple unstructured events and show a rather fine-grained substructure of processes accompanied by pre- and post-conditions. The central point of his argumentation is that although, generally speaking, formal representation describes an action quite well, as exposed by Davidson,[73] various implicit relations between the resulting constituents of the formal expression are neglected. A famous example which shows the relevance of these implicit relations can be found in the above-mentioned Gettier case regarding the formal definition of knowledge. Pross argues that psychological experiments have shown that human beings use implicit strategies such as goal relations and causal structures to comprehend their perceptions in temporal terms. They play a decisive role in understanding behaviour and intention in actions. He proposes a theory of temporal entities which takes temporality into account and thus improves the possibilities of realizing planning, reasoning and representations.

Ludger Jansen's contribution is the final one which deals with knowledge representations. He investigates social entities and, to include actions, aims at extending formal ontologies to the world of social entities. His work is based on Gilbert,[74] Tuomela[75] and Searle,[76] who all deal with the construction of social reality and entities. In order to demonstrate his conception on the basis of a concrete case, Jansen refers to a medical information system. With regard to social entities, it is often difficult to represent the respective object in an adequate way. This often leads to shortcomings such as cultural bias in terminology. Notions always reflect a context-specific perspective, leading to a confusion of universals and particulars, a mix-up of ontological categories, and deficiencies in reflecting the ontic structure of the social world. To tackle these problems, he proposes introducing four classification rules in order to better align ontologies of social events with general ontologies using BFO (Basic Formal Ontology) and OBO (Open Biological Ontologies) standards. His work reflects the specific challenges that we face regarding the representation of social events and actions. Here, we face the same subtleties that we have described with respect to the previous contributions. The only difference is that, in this case, we deal with the abstraction of action, whereas the previous contributions dealt with concrete actions – a difference which has decisive consequences.

The last two sections of contributions leave the area of ontologies and knowledge representations and turn to computation in general. Here, we address the question to which degree the world as a whole can be understood in terms of computation and information. It concerns the questions of whether and how we could replace the traditional matter/energy model, which we know from physics, by a model which is based on the concepts of information and computation to describe the static and dynamic aspects of the universe.[77] This view is called info-computationalism. We can conceive of this approach as a generalization of the analogy between mind and computational process to all processes in world as a whole. In such an interpretation, successful natural intelligent agents are involved in an evolutionary historical process that stands in a multitude of info-computational relations to their environment, including other agents. In order to address the requirements of representing complex systems and explaining emergence, computation must be understood in a sense that goes beyond today's conceptions and includes multiple agents or swarm intelligence. In her contribution, Gordana Dodig-Crnkovic explains the consequences of info-computationalism for information science. She argues that information technology must consider the natural information processes of natural organisms as a template for new technologies. To this end, she analyses current concepts of knowledge and science from an info-computational point of view by following Chaitin,[78] who regards information compression as the most important feature of science. Accordingly, science mainly appears as a tool of sense-making, whereas the certainty suggested by natural sciences does not exist. In info-computationalism, life and intelligence can act autonomously and store (learn), retrieve (remember) information and anticipate future events in their environment. Dodig-Crnkovic concludes her description of info-computationalism with some remarks concerning usual misinterpretations. For instance, the approach does not compare the human mind to a computer. This would leave human beings without any free will, since it would mean that their behaviour is determined by a fixed programme. Info-computationalism describes computational processes beyond such deterministic schemas, in the same way as quantum mechanics goes beyond the deterministic schema of Newtonian mechanics. The central idea of this approach is not a simplification of existing theories, but a reinterpretation of their meaning.

Vincent Müller responds to Dodig-Crnkovic's info-computational approach, offering an analysis of the underlying concept of computation. He explains three viewpoints about pancomputationalism, as a slightly weaker variant of info-computationalism: (a) the view that any future state of an object can be *described* as a computational result from its present state; (b) the view that any future state of an object can be *explained* as computational result starting from its present state; (c) the view that the future state of an object *is* the computational result of its current state. He further introduces a distinction of realist and anti-realist info-computationalism and argues that the anti-realist position is not consistent with the general approach, so that he concentrates on the realist version. He analyses position (b) which he summarizes with the motto 'the universe is a computer'. However, the attempt to reduce physical to computational processes does not appear to be feasible to him since, as he argues, 'computation is not constrained enough to explain physical reality'. His conclusion regarding statement (c) is that 'a complete theory of the universe can be formulated in computational terms'. Here, computational description turns into simulation. Müller relates his concerns to Putnam, asserting that there are usually several possible formal descriptions for an object or process, so that it would not be possible to pick one of them as the basis for the approach.[79] Finally, he concludes that only the metaphorical use of (a) which is associated with the motto 'the universe can often usefully be described as computational' is acceptable. It is this reading that he considers to be promising, as long as we avoid overstretching its range of validity.

Novel approaches such as info-computationalism might not yet be fully developed, but their new perspectives towards reality are often inspiring. We have to consider this approach together with the problem of complexity, which shows that new structures can emerge in systems that we assume to understand due to the mathematical representation of their dynamic processes. However, such emergence is often unpredictable. Info-computationalism makes us aware that the current deterministic machines face a natural limit of creativity. The main question regarding approaches such as info-computationalism is not whether it is right or wrong, but what it tells us about the world in which we live. Although such a process of clarification might take some time, it will lead us to a better understanding of the world and of ourselves.

The concluding contribution by Francis Dane is concerned with the necessity of ethical codes which help computer scientists to decide whether their behaviour complies with the generally accepted norms. The fact that the two leading associations of computer scientist in the USA, the Association for Computing Machinery (ACM) and the Institute of Electrical and Electronics Engineers (IEEE), have fostered such a code indicates that it seems to be an urgent issue. Dane explains that for obvious reasons, such code cannot determine all cases of ethical issues and requires additional ethical competence that, for example, might be acquired by specific training and is centrally based on philosophical experience. In this respect, he refers to Aristotle, Jeremy Bentham, William James and John Stuart Mill as protagonists of discussions regarding the public interest, a discussion that has also become relevant for computer science and its applications. He reminds us of Kant's autonomy principle as the basis for human dignity. However, he also examines sources for the description of ethical behaviour in the work of twentieth-century philosophers such John Rawls, Martha Nussbaum and Amartya Sen. Understanding these issues also requires a technical understanding of the consequences of the application of computers and other machines. In this respect, computer science and philosophy have to go hand in hand.

If we survey the contributions to this essay volume, it is obvious that they address quite independent topics. However, we find various connections under the surface. The central axis consists in the relation between formal representations, which provide the abstract basis of computation, and the (inter)action of human beings with the machine and with each other. They point at a fundamental duality of static objective and dynamic cultural and contextual dimensions of information that are to be reflected in information science. The large number of definitions of information, including their variances, reflects this. It is not possible to neglect one side in favour of another. The contributions also show that the underlying duality is not yet fully understood, in particular since there is another polarity, namely that between traditional philosophers and industrial practitioners. They describe the same problem in different terms. This project should foster the mutual understanding of both groups and encourage steps towards a continuous dialogue between them, even though both sides still hesitate to talk to each other. The attempt to understand the others' language will provide a basis for a successful dialogue, which is not only useful, but even necessary. A philosophy which inspires practitioners is also a motivation for philosophers, and practitioners will listen to philosophers more openly if they realize that philosophy gives them valuable insights. We can achieve this in a direct dialogue between philosophers and practitioners. The conference *Philosophy's Relevance in Information Science* and the resulting book is eager to start and to intensify such dialogue.

1 THE FOURTH REVOLUTION IN OUR SELF-UNDERSTANDING

Luciano Floridi

The First Three Revolutions

To oversimplify, science has two fundamental ways of changing our understanding. One may be called *extrovert*, or about the world, and the other *introvert*, or about ourselves. Three scientific revolutions in the past had great impact both extrovertly and introvertly. In changing our understanding of the external world, they also modified our conception of who we are, that is, our self-understanding. The story is well known, so I shall recount it rather quickly.

We used to think that we were at the centre of the universe, nicely placed there by a creator God. It was a most comfortable and reassuring position to hold. In 1543, Nicolaus Copernicus published his treatise on the movements of planets around the sun. It was entitled *On the Revolutions of Celestial Bodies* (*De Revolutionibus Orbium Coelestium*). Copernicus probably did not mean to start a 'revolution' in our self-understanding as well. Nonetheless, his heliocentric cosmology forever displaced the earth from the centre of the universe and made us reconsider, quite literally, our own place and role in it. It caused such a profound change in our views of the universe that the word 'revolution' begun to be associated with radical scientific transformation.

We have been dealing with the consequences of the Copernican revolution since its occurrence. Indeed, it is often remarked that one of the significant achievements of our space explorations has been a matter of external and comprehensive reflection on our human condition. Such explorations have enabled us to see earth and its inhabitants as a small and fragile planet, from outside. Of course, this was possible only thanks to information and communication technologies (ICTs). Figure 1.1 reproduces what is probably the very first picture of our planet, taken by the US satellite *Explorer VI* on 14 August 1959.

Figure 1.1: First picture of earth, taken by the US satellite *Explorer VI*. It shows a sunlit area of the Central Pacific Ocean and its cloud cover. The signals were sent to the South Point, Hawaii tracking station, when the satellite was crossing Mexico. Image courtesy of NASA, image number 59-EX-16A-VI, date 14 August 1959.

After the Copernican revolution, we retreated by holding on to the belief in our centrality, at least on planet Earth. The second revolution occurred in 1859, when Charles Darwin (1809–82) published his *On the Origin of Species by Means of Natural Selection, or the Preservation of Favoured Races in the Struggle for Life*. In his work, Darwin showed that all species of life have evolved over time from common ancestors through natural selection. This time, it was the word 'evolution' that acquired a new meaning.

The new scientific findings displaced us from the centre of the biological kingdom. As had been the case with the Copernican revolution, many people did not like it. Indeed, some people still resist the very idea of evolution, especially on religious grounds. But most of us have moved on, and consoled ourselves with a different kind of importance and a renewed central role in a different space, one concerning our mental life.

We thought that, although we were no longer at the centre of the universe or of the animal kingdom, we were still the masters of our own mental contents, the species completely in charge of its own thoughts. This defence of our centrality in the space of consciousness came to be dated, retroactively and simplistically, to the work of René Descartes. His famous 'I think therefore I am' could be interpreted as also meaning that our special place in the universe had to be identified not astronomically or biologically but mentally, with our ability of conscious self-reflection, fully transparent to, and in control of, itself. Despite Copernicus and Darwin, we could still regroup behind a Cartesian trench. There, we could boast that we had clear and complete access to our mental contents, from ideas to motivations, from emotions to beliefs. Psychologists thought that introspection was a sort of internal voyage of discovery of mental spaces. William James still considered introspection a reliable, scientific methodology. The mind was like a box: all you needed to do to know its contents was to look inside.

It was Sigmund Freud who shattered this illusion through his psychoanalytic work. It was a third revolution. He argued that the mind is also unconscious and subject to defence mechanisms such as repression. Nowadays, we acknowledge that much of what we do is unconscious, and the conscious mind frequently constructs reasoned narratives to justify our actions afterwards. We know that we cannot check the contents of our minds in the same way we search the contents of our hard disks. We have been displaced from the centre of the realm of pure and transparent consciousness. We acknowledge being opaque to ourselves.

There are now serious doubts about psychoanalysis as a scientific enterprise, and yet one may still be willing to concede that, culturally, Freud was very influential in initiating the radical displacement from our Cartesian certainties. What we mean by 'consciousness' has never been the same after Freud, but we may owe him more philosophically than scientifically. If so, then one could replace psychoanalysis with contemporary neuroscience as a likely candidate for such a revolutionary scientific role. Either way, the result is that today we admit that we are not immobile, at the centre of the universe (Copernican revolution), that we are not unnaturally separate from the rest of the animal kingdom (Darwinian revolution), and that we are very far from being Cartesian minds entirely transparent to ourselves (Freudian or neuroscientific revolution).

One may easily question the value of the interpretation of these three revolutions in our self-understanding. After all, Freud himself was the first to read them as part of a single process of gradual reassessment of human nature. His interpretation was, admittedly, rather self-serving. Yet the line of reasoning does strike a plausible note, and it can be rather helpful to understand the information revolution in a similar vein. When nowadays we perceive that something very significant and profound is happening to human life, I would argue that our intuition is once again perceptive, because we are experiencing what may be described as a fourth revolution, in the process of dislocation and reassessment of our fundamental nature and role in the universe.

The Fourth Revolution

After the three revolutions, we were left with our intelligent behaviour as the new line of defence of our uniqueness. Our special place in the universe was not a matter of astronomy, biology or mental transparency, but of superior thinking abilities. As Blaise Pascal poetically puts it in a famous quote: 'man is but a reed, the most feeble thing in nature, but he is a thinking reed'.

Intelligence was, and still is, a rather vague property, difficult to define, but we were confident that no other creature on earth could outsmart us. Whenever a task required intelligence, we were the best by far, and could only compete with each other. We thought that animals were stupid, that we were clever, and this seemed the reassuring end of the story, even if Thomas Hobbes had already argued that thinking was nothing more than computing ('reckoning') with words. We quietly presumed to be at the centre of the space represented by intelligent behaviour.

The result was that we both misinterpreted our intelligence and underestimated its power. We had not considered the possibility that we may be able to engineer autonomous machines that could be better than us at processing information logically and were therefore behaviourally smarter than us whenever information processing was all that was required to accomplish an otherwise intelligent task. The mistake became clear in the work of Alan Turing, the father of the fourth revolution.

Because of Turing's legacy, today we have been displaced from the privileged and unique position we thought we held in the realm of logical reasoning and corresponding smart behaviour. We are not the undisputed masters of the infosphere. Our digital devices carry out more and more tasks that would require intelligence if we were in charge. We have been forced to abandon once again a position that we thought was 'unique'. The history of the word 'computer' is indicative of this. Between the seventeenth and the nineteenth century, it was synonymous with 'a person who performs calculations', simply because there was nothing else in the universe that could compute. In 1890, for example, a competitive examination for the position of 'computer' in the US Civil Service had sections on 'orthography, penmanship, copying, letter-writing, algebra, geometry, logarithms, and trigonometry'.[1] It was still Hobbes's idea of thinking as reckoning. Yet by the time Turing published his classic paper entitled *Computing Machinery and Intelligence*, he had to specify that in some cases he was talking about a '*human* computer', because by 1950 he knew that 'computer' no longer referred only to a person who computes. After him, 'computer' entirely lost its anthropological meaning and of course became synonymous with a general purpose, programmable machine, what we now call a Turing machine.

After Turing's groundbreaking work, computer science and the related ICTs have exercised both an extrovert and an introvert influence on our understanding. They have provided unprecedented scientific and engineering powers over natural and artificial realities. And they have cast new light on who we are, how we are related to the world and to each other, and hence how we conceive ourselves. Like the previous three revolutions, the fourth revolution has not only removed a misconception about our uniqueness, it has also provided the conceptual means to revise our self-understanding in new terms. We are slowly accepting the post-Turing idea that we are not Newtonian, standalone and unique agents, some Robinson Crusoe on an island, but rather informational organisms (inforgs), mutually connected and embedded in an informational environment, the infosphere, which we share with other informational agents, both natural and artificial, that can process information logically and autonomously often better than we do. Such agents are not intelligent like us, but they can easily outsmart us.

Inforgs

We have seen that we are probably the last generation to experience a clear difference between online and offline environments. Some people already spend most of their time online. Some societies are already hyperhistorical. If home is where your data are, you probably already live on Google Earth and in the cloud. Artificial and hybrid (multi)agents, i.e. partly artificial and partly human (consider, for example, a bank) already interact as digital agents with digital environments, and since they share the same nature they can operate within them with much more freedom and control. We are increasingly delegating or outsourcing our memories, decisions, routine tasks and other activities to artificial agents in ways that will be progressively integrated with us and with our understanding of what it means to be a smart agent. Yet all this is rather well known. And although it is relevant to understanding the displacement caused by the fourth revolution, namely what we are not uniquely, it is not what I am referring to when talking about inforgs, that is, what the fourth revolution invites us to think we may be. Indeed, there are at least three more potential misunderstandings against which the reader should be warned.

First, the fourth revolution concerns, negatively, our newly lost 'uniqueness' (we are no longer at the centre of the infosphere) and, positively, our new way of understanding ourselves as inforgs. The fourth revolution should not be confused with the vision of a 'cyborged' humanity. This is science fiction. Walking around with something like a Bluetooth wireless headset implanted in your ear does not seem the best way forward, not least because it contradicts the social message it is also meant to be sending: being on call $^{24}/_{7}$ is a form of slavery, and anyone so busy and important should have a personal assistant instead. A similar reasoning could be applied to other wearable devices, including Google Glasses. The truth is rather that being a sort of cyborg is not what people will embrace, but what they will try to avoid, unless it is inevitable.

Second, when interpreting ourselves as informational organisms I am not referring to the widespread phenomenon of 'mental outsourcing' and integration with our daily technologies. Of course, we are increasingly dependent on a variety of devices for our daily tasks, and this is interesting. However, the view according to which devices, tools and other environmental supports or props may be enrolled as proper parts of our 'extended minds' is still based on a Cartesian agent, who is stand-alone and fully in charge of the cognitive environment, which it is controlling and using through its mental prostheses, from paper and pencil to a smartphone, from a diary to a tablet, from a knot in the handkerchief to a computer. This is an outdated perspective.

Finally, I am not referring to a genetically modified humanity, in charge of its informational DNA and hence of its future embodiments. This posthumanism, once purged of its most fanciful and fictional claims, is something that we may see in the future, but it is not here yet, either technically (safely doable) or ethically (morally acceptable). It is a futuristic perspective.

What I have in mind is rather a quieter, less sensational and yet more crucial and profound change in our conception of what it means to be human. We have begun to understand ourselves as inforgs not through some biotechnological transformations in our bodies, but, more seriously and realistically, through the radical transformation of our environment and the agents operating within it.

Enhancing, Augmenting and Re-Engineering Technologies

The fourth revolution has brought to light the intrinsically informational nature of human identity. It is humbling, because we share such a nature with some of the smartest of our own artefacts. Whatever defines us, it can no longer be playing chess, checking the spelling of a document, calculating the orbit of a satellite, parking a car or landing an aircraft better than some ICT. And it is enlightening, because it enables us to understand ourselves better as a special kind of informational organism. This is not equivalent to saying that we have digital alter egos, some Messrs Hydes represented by their @s, blogs, tweets or https. This trivial point only encourages us to mistake ICTs for merely *enhancing* technologies, with us still at the centre of the infosphere. Our informational nature should not be confused with a 'data shadow' either, an otherwise useful term introduced to describe a digital profile generated from data concerning a user's habits online. The change is deeper. To understand it, consider the distinction between *enhancing* and *augmenting* technologies.

The switches and dials of enhancing technologies are interfaces meant to plug the appliance into the user's body ergonomically. Axes, guns and drills are perfect examples. It is also the cyborg idea. The data and control panels of augmenting technologies are instead interfaces between different possible worlds. On the one hand, there is the human user's outer world as it affects the agent inhabiting it. On the other hand, there is the dynamic, watery, soapy, hot and dark world of the dishwasher, for example. Or the equally watery, soapy, hot and dark but also spinning world of the washing machine. Or the still, aseptic, soapless, cold and potentially luminous world of the refrigerator. These robots can be successful because they have their environments 'wrapped' and tailored around their capacities, not vice versa. Now, despite some superficial appearances, ICTs are not enhancing nor augmenting in the sense just explained. They are environmental forces because they are creating and re-engineering whole realities that the user is then enabled to enter through (possibly friendly) gateways. It is a form of initiation.

Looking at the history of the mouse, for example, one discovers that our technology has not only adapted to, but also educated, us as users. Douglas Engelbart, the inventor of the mouse, once told me that he had experimented with a mouse to be placed under the desk, to be operated with one's knee, in order to leave the user's hands free. After all, we were coming from a past in which typewriters could be used more successfully by relying on both hands. Luckily, the story of the mouse did not go the same way as the story of the QWERTY keyboard, which never overcame the initial constraints imposed by the old typewriters. Today, we just expect to be able to touch the screen directly. HCI (to remind you: human–computer interaction) is a symmetric relation. Or consider current attempts to eliminate screens in favour of bodily projections, so that you may dial a telephone number by using a virtual keyboard appearing on the palm of your hand. No matter how realistic this may be, it is not what I mean by referring to the development of inforgs. Imagine instead the current possibility of dialling a number by merely vocalizing it because your phone 'understands' you.

To return to the initial distinction, while a dishwasher interface is a panel through which the machine enters into the user's world, a digital interface is a gate through which a user can be present[2] in the infosphere. This simple but fundamental difference underlies the many spatial metaphors of 'cyberspace', 'virtual reality', 'being online', 'surfing the web', 'gateway' and so forth. It follows that we are witnessing an epochal, unprecedented migration of humanity from its Newtonian, physical space to the infosphere itself as its new environment, not least because the latter is absorbing the former. As a result, humans will be inforgs among other (possibly artificial) inforgs and agents operating in an environment that is friendlier to informational creatures. And as digital immigrants like Generation X and Y are replaced by digital natives like Generation Z, the latter will come to recognize no fundamental difference between the infosphere and the physical world, only a change in perspective. When the migration is complete, my guess is that new generations will increasingly feel deprived, excluded, handicapped or poor to the point of paralysis and psychological trauma whenever it is disconnected from the infosphere, like fish out of water. One day, being an inforg will be so natural that any disruption in our normal flow of information will make us sick.

Conclusion: Digital Souls, their Value and Protection

In the light of the fourth revolution, we understand ourselves as informational organisms among others. In the long run, *de-individualized* (you become 'a kind of') and *re-identified* (you are seen as a specific crossing point of many 'kinds of') inforgs may be treated like commodities that can be sold and bought on the market of advertisements. We may become like Gogol's *dead souls*, but with wallets.[3] Our value may depend on our purchasing power as members of a customer set, and the latter is only a click away. In a way, this is all very egalitarian: nobody cares who you are on the web, as long as your ID is that of the right kind of shopper.

There is no stock exchange for these dead souls online, but plenty of Chichikovs (the main character in Gogol's novel) who wish to buy them. So what may an inforg be worth, in dollars? As usual, if you buy them in large quantities you get a discount. So let's have a look at the wholesale market. In 2007, Fox Interactive Media signed a deal with Google to install the famous search engine (and related advertising system) across its network of internet sites, including the (at the time) highly popular MySpace. Cost of the operation: $900 million.[4] Estimated number of user profiles in MySpace: nearly 100 million at the time of the deal. So, average value of a digital soul: $9 at most, but only if it fitted the high-quality profile of a MySpace.com user. As Sobakievich would say: 'It's cheap at the price. A rogue would cheat you, sell you some worthless rubbish instead of souls, but mine are as juicy as ripe nuts, all picked, they are all either craftsmen or sturdy peasants.'[5] The 'ripe nuts' are what really counts, and in MySpace, they were simply self-picked: tens of millions of educated people, with enough time on their hands (they would not be there otherwise), sufficiently well-off, English-speaking, with credit cards and addresses in deliverable places ... it makes any advertiser salivate. Fast-forward five years, and the market is even bigger, the nuts much less ripe, so prices even lower. In 2012, Facebook filed for a $5 billion initial public offering.[6] Divide that by its approximately 1 billion users at that time, and you have a price of $5 per digital soul. An almost 50 per cent discount. You can imagine my surprise when, in 2013, Yahoo bought Tumblr (a blogging platform) for $1.1 billion: with 100 million users, that was $11 per digital soul. I suspect (at the time of writing: Monday 13 June 2013) it might have been overpriced.[7] Consider that, according to the *Financial Times*,[8] in 2013 most people's profile information (an aggregate of age, gender, employment history, personal ailments, credit scores, income details, shopping history, locations, entertainment choices, address and so forth) sold for less than a dollar in total per person. For example, income details and shopping histories sold for $ 0.001 each. The price of a single record drops even further for bulk buyers. When I ran the online calculator offered by the *Financial Times*, the simulation indicated that 'marketers would pay approximately for your data: $0.3723'. As a digital soul, in 2013, I was worth about a third of the price of a song on iTunes.

From Gogol to Google, a *personalizing* reaction to such massive *customization* is natural but also tricky. We construct, self-brand and reappropriate ourselves in the infosphere by using blogs and Facebook entries, Google homepages, YouTube videos and Flickr albums, by sharing choices of food, shoes, pets, places we visit or like, types of holidays we take and cars we drive, instagrams, and so forth, by rating and ranking anything and everything we click into. It is perfectly reasonable that *Second Life* should be a paradise for fashion enthusiasts of all kinds. Not only does it provide a new and flexible platform for designers and creative artists, it is also the right context in which digital souls (avatars) intensely feel the pressure to obtain visible signs of self-identity and unique personal tastes. After all, your free avatar looks like anybody else's. Years after the launch of *Second Life*, there is still no inconsistency between a society so concerned about privacy rights and the success of social services such as Facebook. We use and expose information about ourselves to become less informationally anonymous and indiscernible. We wish to maintain a high level of informational privacy almost as if that were the only way of saving a precious capital that can then be publicly invested (squandered, pessimists would say) by us in order to construct ourselves as individuals easily discernible and uniquely reidentifiable by others.

2 INFORMATION TRANSFER AS A METAPHOR

Jakob Krebs

Introduction

An intuitive understanding of information concerns the means by which knowledge is acquired. This notion corresponds to an instantiation of the complementary properties of 'being informative' and 'being informed'. It is in this intuitive sense that the term 'information' is used in many accounts of verbal communication and explicit learning. But the term is furthermore prominent in accounts of genetics, neurobiology and cognitive science, as well as in communications engineering, and computer sciences. In view of this transdisciplinary use of the word 'information', some theoreticians hope for a unified conception that would serve as a common denominator in interdisciplinary investigations.[1] According to this promise, we would then be equipped with a singular conceptional grasp on physical and genetic structures, on neuronal patterns as well as on cognitive and communicative events. Unfortunately, a unified concept of information is far from being spelled out, since many of the disciplines mentioned above follow quite different approaches.[2] Even in the context of 'information science' itself, extensive differences prevail on the notions of data, information and knowledge and their conceptual interconnections.[3] But if we detect not a single but various conceptions of information, we should not expect a single but various theories of information – a point made by Claude Shannon long before the transdisciplinary implementation of his mathematical theory.[4] When one or the other 'theory of information' gets implemented into theories of communication or learning, for example, these theories thereby inherit one particular conception of information. Likewise, any understanding of 'the' information society or 'the' information science is already impregnated with an implicit or explicit conception of information.

The following metaphorological analysis challenges explanations in terms of *transferable information*, when they claim to describe, understand or model processes of verbal communication and explicit learning. It starts with an assessment of the very idea of transferability and an outline of related philosophical debates (pp. 30–1). The second section uncovers the metaphoric implications of the assumed information transfer in the light of contextualist perspectives (pp. 32–5). In order to illustrate the downside of metaphorical explanations, a similarly misleading metaphor is discussed with general remarks on metaphors in science in the third section (pp. 36–7). A concluding prospect on the relevance of philosophical and especially metaphorological inquiries in information science complete the chapter (pp. 37–40).

Transferability and Context

It is noteworthy that the use of the term 'information' is very often conceptually associated with a 'transmission', a 'flow' or a 'conveyance'. In fact, the very idea of the transferability of information suggests a promising explanatory potential, since the assumed transfer implies a related causal force or at least a structural reproduction. With regard to these implications, the explanatory burden is divided between the invested notions of 'information' and 'transfer', when one tries to explain processes of verbal communication or explicit learning as a sort of 'transfer of information'. Leaving aside the notion of information for a moment, one should separately examine the explanatory role of the ideas of a 'transfer' or a 'flow' together with associated verbs like 'carrying', 'conveying', 'absorbing' or 'filtering' information. Apparently, all those ideas imply the movement of some entity from one location to another. If this idea features as part of an explanans, the resulting explanations are committed to the covering of a distance by some mode of travelling. By claiming that communication or learning can be explicated as a transfer of information, the idea of a transfer is complemented by designating the moving good as 'information'. The thrust of explanations with 'information' as an objective commodity draws more or less explicitly on the idea of its *transferability*. An example is when the communication of knowledge is modelled as the 'gathering of pieces' of information, thereafter 'stored' somewhere in the brain. In view of the formal structure of the explanation, a change of location is a vital part of the explanans that is supposed to explain the underlying processes of communication or learning. The appeal of this type of explanation clearly lies in the elegant conceptual bridge it builds between mind and world, with information as something interchangeably residing in both realms.

One can challenge these explicit or implicit explanations in terms of an alleged information transfer as being inapt to account for complex forms of communication or learning, since one can, for example, provide information about different states of affairs with the same expression. So in many contexts of communication and learning, the term 'information' can only refer to expressive or other events as 'being informative' in certain respects. Being informative is not a transferable entity, but a relational property, which supports no direct implications on causal power or structural reproduction. The assumed independent ontological status of 'transferred information' can be contested,[5] since 'information' in the context of verbal communication or explicit learning can be identified only in the form of inferentially embedded propositional content. This holistic, inferential or contextual aspect of informativeness is strictly incompatible with the notion of transferability and the causal or functional explanations it promises. With these two disparate conceptions behind the term 'information', one should not expect to find any insights into the 'nature of information', regardless of the theoretical differences the vocable 'information' conceals.

In philosophy, the term 'information' often occurs at the interface of epistemology, semantics and the philosophy of language, which does not mean that it is uncontroversial. Most importantly, the scientific community discusses how to relate *natural* conceptions of information to *intentional* ones, where the latter are reserved for conceptual communication and learning.[6] For example, at the one end of the spectrum we find approaches towards a naturalized epistemology, which use Shannon's statistical conception of information[7] in order to explain knowledge as the outcome of a flow of information – the latter presented as an 'objective commodity'.[8] At the other end of the spectrum we find the view that the term 'information' primarily refers to a kind of expressive content, which cannot be objectively located or transferred.[9] As Davidson states, information 'as we know and conceive it has a propositional content geared to situations, objects, and events we can describe in our homespun terms'.[10] Connected to the analogous difference between knowledge by perception and knowledge by testimony is the ongoing debate on the nature of non-conceptional content and its relation to the conceptual realm.[11] Related issues concerning justification by testimony and contextually varying ascriptions of knowledge are also central to debates on philosophical contextualism:[12] 'information' can count as the content of propositions irrespective of a given context or it can refer to the content of utterances modulated in accordance with contextually varying pragmatic principles.[13] The doubts about the idea of an information transfer rest on the fact that at least *some* of the semantic properties of an expression are relational properties. Even Jerry Fodor, although well-known for his sceptical assessment of semantic holism, declares that the content of a sentence is potentially 'sensitive to which belief systems it's embedded in'.[14] Since those conceptual tensions cannot readily be resolved, an informational unification is by no means without problems.[15] Far from being settled, conceptions of 'information' are differentiated and recontoured in the light of the different theoretical approaches and their respective explananda. Ultimately, philosophers discuss the ontological status of information, as it is understood and presented by the different epistemologic, semantic or pragmatic accounts.

The Ontological Status of Information

Relational intuitions in general are concerned with the idea that semantic properties of expressions or concepts do not persist in isolation, but only in relation to other expressions or concepts. According to a holistic perspective, the content of expressions is 'ontologically dependent in a generic way'[16] on their actual arrangement with other expressions or concepts. An influential version of this view originates from Davidson, who initially maintains the principle of compositionality, according to which the meaning of a sentence can be derived from the meaning of its parts. But he claims *additionally* that the meaning of the parts depend on the roles they play in complex expressions.[17] Especially in the late adjustments of his theory of interpretation, he points to intentional and social factors, which guide the reconstruction of the content of *uttered* propositions. The illocutionary force of expressions is thereby related to other acts with respect to practical concerns. Moderate versions of a semantic holism must not claim that *all* other propositions play a role, but that the content of a proposition might depend on a subset of related propositions in play.[18] For example, to understand what it means for some object to be red, one doesn't have to get one's mind around *all* the beliefs or concepts one holds, but one can have a full conception of something being red only by knowing that red is a colour, which means to be one colour among others. Even in such a moderate account of semantic holism, for someone to be informed about something being red, a mere causal contact with red objects in the past will not suffice.

This emphasis on the relational properties of *informative expressions* is neither a radical constructivist perspective nor a purely linguistic remark. On the contrary, it fosters the idea that our conception of content cannot be developed without taking the interconnectedness of worldly, practical and linguistic elements into account, an idea that is presented in Wittgenstein's relation of language-games to forms of life.[19] In this line of thought, our access to the world, the social practices we are engaged in and our linguistic capacities must be understood as mutually dependent factors.[20] In order to become informed about one's bus leaving at 15:30, one need not only know what buses are, but also what '15:30' denotes in this context, and, furthermore, what it means for a bus to leave in accordance with a time schedule. Likewise, when we long for information about economy, politics or sports, we will find answers to our questions only in respect to our preceding knowledge – when '15:30' informs about scores in tennis, for example. In order to learn about those complex states of affairs, it is not sufficient to 'gather' or to 'store' information in the form of raw data or true sentences. Parrots or computers might reproduce syntactical or phonological features of those without any idea what the truth of sentences is all about. An interpreter must instead engage in an active reconstruction of the phenomena and relations an uttered expression is *used to inform about*.

Consider, for example, the sentence 'I am now here'. As far as semantic minimalism is concerned, its content is governed by truth conditions despite the indexicals involved.[21] In this special case, the truth conditions are satisfied for every utterance, while the sentence expresses the same meaning across all contexts. If one is tempted to link the informational value to this minimal semantic meaning of the sentence, one will find it utterly uninformative, since it is always true if uttered (although not necessarily true in every possible world).[22] From the perspective of pragmatically motivated versions of contextualism, it is no surprise how utterances of this sentence can nonetheless become *informative*: the occasional meaning of its expression does not only depend on the contextual fixation of the involved indexicals, but also on additional assumptions of intentions, locations and further coordinative respects in a practical context. If you know that I took the train to Stockholm and I send an email solely stating 'I am here now', this expression is informative, given you kept track of my plans – something I should assume at the time I write the line. This example nicely shows how the property of being informative is not linked to truth or truthfulness in a straightforward way, but is brought out 'by adding in collateral knowledge'.[23] More controversially with regard to truth conditions, the sentence 'I am not here' appears to be false for any context of utterance. Still, it can become informative if expressed in order to prevent a third person knowing about one's location. It is a case of a conventional implicature in the context of undesired phone calls and has an informative character, although its truth conditions can hardly be met by any utterer.

As promoted by the idea of speech act pluralism, an utterance can assert, say or claim propositions that are not (logically) implied or that are even incompatible with the 'proposition semantically expressed by that utterance'.[24] 'Semantically' here refers to the truth conditions of the sentence in the first place, but since an *uttered* sentence can instantiate different occasional meanings, it is a continuous philosophical issue where to draw the distinction between semantics and pragmatics when it comes to the fixation of meaning in a given context.[25] Since assertions are often characterized with respect to their informative value, the issue on meaning-fixation is closely related to the question of how to account for this informational character. To complicate this outline of some philosophical debates one step further, some approaches try to understand context itself 'as a kind of informational state'.[26] From this perspective, 'information' also refers to the set of propositions which are common ground in a given conversation. To present an assertion is thereby understood as the task of updating a context, which is supposed to recast the idea of an assertion 'conveying' information only against a 'background of shared information'.[27] But if those views take it to be sufficient for the participants of a conversation to take something for granted, while allowing that they do not in fact believe it,[28] the intuitive connection of information and knowledge seems to be undermined.

Especially if one is sympathetic towards an account of semantic minimalism, explications of the vocable 'information' as a factor of successful communication should be expected on the level of speech acts. That is because *informativeness* appears to be a feature of utterances in the first place, not of sentences devoid of their contextual application. Expressions become meaningful – and eventually informative – predominantly with respect to our acquaintance with epistemic social practices, where an interpreter must modulate and enrich an expression in order to infer the most plausible interpretations.[29] Moreover, informativeness is not at all an exclusive property of semantic phenomena. Practically any event can become informative in dependence of knowledge, interests and investigative competences of an interpreter in a given context. Regarding verbal communication and explicit learning, we are confronted with different formats of expressions and their usage in models and explanations for a wide variety of phenomena. This relational understanding of information thereby stands in marked contrast to the notions of information used in the psychology of perception or the one of telecommunication engineering. In the latter cases 'information' refers to something like regular, wavelike patterns that reproduce over a number of transductions, being somehow functionally effective. Doubts have been raised in this regard not only about linear models of transferable *genetic* information[30] but also about models of the brain as a *neuronal* information processor in the philosophy of cognitive science.[31] Concerning theories of communication and learning, the idea of causal chains can hardly deal with the fact that the very same linguistic pattern can have different informational value on different occasions of an utterance.

It is not as if contextualism provides a similarly straightforward explanation of content or information. First of all, it stresses the finding that content cannot be reduced to transferable information of some sort, since it does not necessarily supervene on the (syntactic) structure of events. Fodor, equally precautious towards hasty approaches of computational psychology and semantic holism, explains the seductive conceptual identification of two senses of information with a confusion about the term of *representation*: it is one thing to conceive of 'content' as the outcome of a causal relation between world and mind, where a given stimulus constitutes a mental label for its source. But 'content' means something different when we think about absent entities, when we model possible worlds or when we learn from others by communicative means – then we 'represent' the phenomena in question without direct causal connection.[32] In the latter cases it is not the direct contact to material entities (that may well play a role for the content of a perception), but the interrelations between one's thoughts that endow the content. For example, thoughts about abstract or unacquainted entities like centres of gravity or black holes derive their content from an interconnection of antecedent thoughts, never from causal contact. Considering furthermore our ability to be informed via implicit hints, vague terms, indexicals, irony or metaphors, theories of verbal communication or explicit learning are forced to draw on inferentially oriented conceptions of information. In order to address the far-reaching questions on communication, we cannot simply ignore the inferential dimensions of *understanding*. In regard of the reservations of contextualists, it is far from clear how communicative and interpretative *acts* should be explicated in terms of a transfer of information. The same holds for approaches to explicit learning, since its complementary predicate 'to inform someone' clearly separates explicit comprehension from simple labelling or conditioning.

In explanations of communication and learning, the idea of transferability appears to be an ontologically misleading metaphor, since explicit information can be individuated solely in propositional form, which derives its occasional meaning with respect to the context and the epistemic background of an interpreter. The idea of a transferable, objective commodity is thereby contested with regard to an interpreter for whom all kinds of events (not only linguistic ones) can become informative in different respects. From the perspective of contextualism, the transferability clause is incapable of fulfilling its promised explanatory job. Instead it is marked as a metaphor that cannot be explicated in relational terms since it cannot be reconciled with the idea of inferentially individuated content. Here the fundamental problem for explanations in terms of transferability arises: if no transferred commodity can be ontologically identified, the explanatory force of the model of transferability is dissolved, relinquishing at least one conception of 'information' to relational definitions. Simply trying to separate any concept of information from the transferability assumption means to change the kind of explanation. This happens, for example, if information is considered to be a result of interpretation in the form of truthful declarative utterances like in a refined 'theory of strongly semantic information'.[33] Since the informative force of assertions depends not only on the epistemic history but also the situation and the interests of an interpreter, this particular conception of 'information' – separated from the transferability clause – is not so much a solution but raises further questions. Once one acknowledges the relational character of information in the context of verbal communication or explicit learning, it appears to be deceptive to recombine it with the idea of transferability, whether one hopes for universal explanations or not. In contrast to explanations associated to transferability as well as purely semantic conceptions, pragmatic theories are needed to account for the ingredients and factors of informative expressions 'that go beyond what is given by the semantics alone'.[34]

Metaphorological Analysis

The critique against the idea of a transfer of information as a misleading metaphor draws on the *relational character of informative expressions*. A few remarks on the metaphorological method used to deconstruct explanations that draw upon the concept of a transfer will further develop the sceptical stance. In the philosophy of science it is not a new finding that metaphors do not serve merely elliptic or ornamental purposes, but can also establish new perspectives on conceptually underdetermined phenomena.[35] Metaphors often serve as a speculative starting point in the search for innovative models and explanations. When a metaphorically gained perspective is deployed in a scientific model, this metaphor 'dies' in the sense that it is not perceived as a metaphor any longer. 'Dead metaphors' are thereby added to the repertoire of meaningful phrases, while being inferentially integrated in the everyday use of language or more specialized discourses.[36] The problem with the use of metaphors as 'intuition pumps' lies in their potential to frame some aspects of the phenomena often only at the cost of ignoring others.[37] Therefore, intuition pumps still need to be carefully inspected, since they come without any guarantee of working properly. In order to rate any metaphorically gained model, one needs to assemble further criteria about the aptness of the metaphorical construction in respect to the conjectured complexity of the phenomena in question. In other words, when used as explanans, metaphoric phrases have to be scrutinized in the light of further criteria for the modelling of the explanandum.

Of course, metaphoric conceptions present in everyday talk are useful in pragmatic respects and serve the function of coordinating the means by which certain effects are reached. They can even provide the sole approach to certain phenomena for some time. Nevertheless, the implicit picture of a metaphorically framed conception might turn out to be rather misleading when one reflects on the structure of the model with respect to the complexity of the modelled phenomenon. As argued above, the idea of the transfer of information as a model for communication or learning can count as such a misleading metaphor. In order to clarify the potentially delusive character of metaphorical conceptions, it is instructive to take a look at an analogous metaphor of transfer from everyday speech: the suggestion that one can 'get rid of a fever by sweating it out' reveals the false assumption that fevers exist 'inside' the body and could be transferred 'outside', with the sweat as a vehicle. While the aim of getting better is often achieved by increasing the body temperature, the metaphorically gained explanation is challenged by contemporary medical knowledge. There simply is no distinct mobile fever-entity inside the body that could be transported outside. 'To get rid of a fever' rather refers to a complex healing process that only appears to be so simple because it is metaphorically modelled as a kind of transportation.

This metaphor can even be held responsible for quite some suffering back in times when the medical practice of bloodletting led to a further weakening of patients instead of transporting 'fevers' outside.

The type of linguistic twist at work here roots in the possibility of reification, whereby a nominalized property ('having a fever') takes the grammatical place of something substantial ('the fever'). Those grammatically generated 'entities' can then be further combined with all kinds of ideas associated with the entities relevant in everyday practice or scientific discourse.[38] In this respect, 'sweating out a fever' appears to be structurally equivalent to the idea of a 'transfer of information' in explanations of communication or learning: an apparent movement of some sort of entity is held responsible for some effect, so it is believed to qualify as part of an explanation for the underlying processes. By looking at the idea of the transfer of information as a misled grammatical reification, the metaphorical nature of the 'sender-receiver-scheme' can be further reconstructed: Just as the ontological implication of 'the fever' rests on the property of 'having a fever', the concept of 'information' refers to nothing substantial in the world but to a relational property of objects or events that can 'be informative' for an interested interpreter. This is not to say that there cannot be objective conceptions of information, since one is of course free to use the term as a referent to objective structures that might be mobile in some sense (neurotransmitter, for example). But when it comes to communicative expressions and explicit learning, structural reproduction and transferability lose their explanatory potential. Thus, we have to look beyond causally effective or 'objective' structures to account for their informational relevance in respect of subjective or intersubjective capacities and interests. In sum, the noun 'information' in contexts of verbal communication and explicit learning refers to a metaphorically reified, relational property, while 'information' conceived of as naked structuredness refers to the intrinsic properties of some event. The causal explanations by means of a transfer depend on the causal effectiveness of intrinsic properties – particularly in the case of computations in a Turing architecture.[39] In contrast, relational properties cannot serve in causal or computational explanations and are incompatible with the idea of a transfer, since their being depends on a relation to a subject. To conceive of information transfer as a metaphor means to adhere to a relational notion of informativeness in contexts of verbal communication and explicit learning, and to reject explanations from transfer because of this.

Outlook

If the argument presented here is sound, ideas of a transfer of information can hardly claim explanatory potential in questions concerning processes of learning or communication. 'Information', explicated as the relational property of 'being informative' and complemented by the state of 'being informed', is itself a term that is in need of an explanation. In the light of this analysis there appear to be at least two mutual exclusive concepts behind the word 'information'. It is not that we have a single concept with different applications but that we have one vocable covering mutually exclusive concepts. For any science of information it is therefore part of its profession to reflect on the conceptual incongruities between information referring to structured entities with causal powers and information as a relational property in the context of epistemic practices. Moreover, if one really wants to predict yet another philosophical turn towards a philosophy of information, one needs further specifications about the conceptions of information in question.[40] Although the conceptual analysis is nowadays not seldom devalued as a kind of unfashionable legacy of dusty armchair-philosophy, the reflection on the ways we conceive our practices in terms of metaphors is by no means out of time. The contemplation of our conceptual relations to the world and each other may be paradigmatic for philosophical endeavours, but every discipline can, did and needs to engage in conceptual reflections.

Since information science is clearly linked to social practices of communication and learning via digital devices, it becomes a pressing issue to relate the differing conceptions of information to each other. For example, bearing the relational property of informativeness in mind, there has never been a society that was not an information society. The assertive mode of speech acts simply is the foremost way by which communities coordinate their practices in accordance to worldly friction.[41] On the other hand, we can scarcely ignore the finding of cultural science that the introduction of digital technology has quite an impact on some of our practices. Nevertheless, for a given digital structure to count as an information artefact we need individuals involved in social practices to make sense of the structures in terms of epistemically conducive evidence. To browse the internet for information does not mean to passively absorb some data, but it means to perform an interested investigation for clues on how one should model diverse dimensions of a shared world. And of course, the more we know about a state of affairs, the easier it is for us to figure out some missing links or actual relations within a rather fixed net of ontological assumptions. So the digital tricks we invent to code and process data must not blind us in regard of the fact that the prevailing medium of the internet remains our contextually flexible language, which we acquire in social settings and by acquaintance with our material surroundings. Information, understood as expressed in propositions linked to truth, is nothing that could be moved – if anything, it is something that can be *shared*, provided that there is a sufficient amount of common background knowledge, geared to viable descriptions of the world.

Without further reflections on disparate conceptions of information, one risks proceeding with confused ideas about our engagement with the world and each other. For example, the idea that theories of learning are first of all subject to neurobiology, which thereby is supposed to shed new light on educational practices, draws heavily on the idea of neurally processed information. But this approach more than often ignores the completely different conception of information in the context of learning within educational practices. Similarly, actual attempts to develop a sustainable notion of media literacy are especially prone to confuse two conceptions of information. On the one hand, media literacy seems to be in need of a conceptual connection to the information and communication technology and engineering problems of information transfer in digital terms. On the other hand, one can hardly want to fall behind the conceptions of communication as an active process of understanding between socialized agents by means of propositional commitment, as developed by philosophers like Habermas or Brandom.[42] Where the latter inferential approaches explicitly draw on skills of interpretation when rendering the act of informing, the technological perspective brings into focus binary data, 'streaming' all over the planet. But since the syntax of binary data has no content in its own right but only in respect to an initiated interpreter, the technical aspects of transmission are mostly marginal to questions about media literacy. Again, the primary medium of knowledge acquisition active in media literacy is our shared language, with the 'content' of mass media being 'informative' first and foremost by propositional means with respect to its relevance for pragmatically embedded subjects with (desirable) interpretative competences. Only from this perspective can one fully appreciate that media literacy, understood as an emancipatory skill, addresses neither mere 'data mining' nor solely the interpretation and reconstruction of informative assertions, but also all sorts of more or less intelligent fictional content that never claims to be informative in the first place.

Therefore, philosophy's relevance in information science derives from its methodological capabilities of reflection, with which the multifarious relations between theories and practices can be assessed. Data is not augmented with meaning to *become information* in the light of a query; this kind of mysterious transsubstantiation is an ontological dead end. Data are just one kind of phenomena that can *become informative*, if they help to satisfy more or less concrete epistemic interests. In order to ask intelligent questions, some background knowledge is already needed in order to limit the range of possibilities, otherwise the given answers will not help in any way. That is why a semantically biased 'erotetic approach'[43] might be a debatable enterprise, as long as it differentiates between information as true semantic content and knowledge by the criteria of questions and understanding. In order for the datum 'twelve' to become informative, the question 'how many apostles were there?' must of course be understood by the investigator in the first place – which means that she needs to already know what apostles are, for example. There are in fact more contrastive explications of knowledge as the capacity to respond in many different ways to questions of many different kinds, which is directed against the oversimplifying distinction between *knowing-that* and *knowing-how*.[44] In accordance with comprehensive epistemological accounts, *to be informative* is the relational property of expressions, objects or events, which feature in the acquisition of knowledge. This reconstruction of *relational informativeness* matches the critique within the information sciences against the misguided idea of transferred information being some transitional state between data and knowledge.[45]

3 WITH ARISTOTLE TOWARDS A DIFFERENTIATED CONCEPT OF INFORMATION?

Uwe Voigt

The Predicament of the 'Concept of Information'

We are talking about 'information' in many different contexts: not only in our ordinary language, but also in the highly specialized discourses of the theory of communication, computer science, physics, biology, cultural studies, and so forth. As we do so, are we using the same concept each and every time? Is there one and only one *concept* of information connecting all these different usages of one word (and its linguistic 'relatives' in languages other than English)? This question has accompanied the 'information talk' for many years[1] and recently has lead to the so-called 'Capurro trilemma'.[2] According to this trilemma, throughout those various contexts the words we use either (A) have the same meaning or (B) completely different meanings or (C) different meanings which nevertheless are somehow connected. As the unity of meaning is a minimal condition for the identity of a concept, in case (A) there is only one concept of information (univocity); in case (B) we deal with several concepts of information (equivocation); in case (C) it is the question of just precisely how the different concepts are interconnected. The authors describing the dilemma suggest Wittgensteinian family resemblance and Aristotelian analogy, but they do not seem to be satisfied by their solutions. Therefore, according to them, we are facing a real trilemma whose single horns are equally unattractive.

At a second glance, the situation of 'information' does not seem to be so bad, though not quite harmless: in many other cases, we are using words to refer to a wide range of entities, and still there is one single concept behind these usages, because all these entities have a common definition. So, locusts, lizards and lions can be defined as animals of a certain kind. In the case of 'information', however, it seems hard, if not impossible, to reach such a definition. S. Ott[3] enumerates eighty different definitions of information, so that the problem leading to the trilemma just reoccurs on this level. And famous instances of definition like Bateson's 'a difference which makes a difference'[4] may seem to be successful just because they are too general, so that they serve as umbrella terms without a clear-cut meaning of their own. Bateson himself confesses that the concept of difference he uses remains quite unclear.[5]

Of course, there might be the one right definition which we just do not know (or have not yet have recognized as such). The basic obstacle for such a definition, however, is the fact that there seem to be mutually exclusive usages of 'information'.[6] On the one hand, 'information' is used to express the process or result of cognitive construction – constructivist concept of information; on the other hand, this very word also serves to refer to objectively given resources which just are out there in the world – objectivist concept of construction. But how can two mutually exclusive concepts be covered by one definition which is not too general ('transcendental' in the medieval sense of the word)? And, given this conceptual diversity, how can there be one science of information, if information is supposed to be a basic concept of this science?

An Aristotelian Proposal

As we have to deal with these questions, a look at Aristotle can be useful. Not that Aristotle himself already had a concept of information; for that, his metaphysical and physical assumptions were too remote from ours.[7] But Aristotle, in his doctrine of the soul, had to handle a problem which was structurally very similar to the predicament depicted here. The soul is defined by Aristotle – according to the beliefs of his time[8] – as the principle of life.[9] Life, however, comes along in many different shapes and sizes and especially in many very different forms and types.[10] There is life as the possession of vegetative capacities like growth and decay, the absorption of nourishment and propagation; there is life as the possession of senso-motorical capacities; there is life as the possession of the capacities for language and discursive thought; and there is – as Aristotle believed – the divine life of pure thought. Aristotle's problem with these concepts of life is: how can there be one concept of life (and, accordingly, of soul)? And how can such a concept apply both to the divine life of pure thought and all other, mortal, forms of life?

Aristotle gains a solution to this problem in two steps. In the first step,[11] he shows that the different types of mortal life and the corresponding concepts of life form an ordered series. For example, there is one lowest, basic type – vegetative life – from which the more complex definitions of the other types are derived. Vegetative life is self-preservation by vegetative acts; different kinds of living beings preserve their bodies by specific acts of this self-preservation. Therefore, in the definitions of the higher types, the basic definition is contained as a determinable element which is determined in different ways. Senso-motorical life, for example, presupposes vegetative capacities of a certain kind, especially those preserving the functioning of the sense organs. Therefore, senso-motorical life is to be defined by reference to these special – nevertheless vegetative – capacities.[12] Even discursive thought presupposes all the other capacities found in mortal living beings, according to Aristotle.[13]

But this first step is not enough. For the divine life of pure thought is *not* defined by any reference to vegetative activities.[14] Hence, the corresponding concept of life is not part of the ordered series. This necessitates the second step: Aristotle's attempt to show that the 'concept of life' has two main meanings which overlap in a certain way. To put it another way: there are indeed two concepts of life, but they are not totally apart from one another. The reason for this claim is that human, discursive thinking is contained in the ordered series of mortal life and at the same time participates in the divine life of pure thought. This participation is expressed by Aristotle in his concept of the 'active mind', our ability of supposing and building theories, which makes possible all our changing cognitive activities, while itself remaining unchanged in the process.[15]

So there are two ordered series: the ordered series of mortal life, culminating in human discursive thinking, and the ordered series of the life of thought, with the basic element of pure divine thought and the derived element of human thinking insofar it is based on the 'active mind'. For each of these series, their common element is a borderline case: participation in pure divine thought for the series of mortal life, participation in mortal life for the series of the life of thought.

In the final analysis, Aristotle therefore is content with having two concepts of life which allows him to have a firm basis also for his empirical studies (the concept of mortal life) without being forced to exclude other – at least thinkable – forms of life.[16]

Application to the Predicament of the 'Concept of Information'

This Aristotelian solution cannot be directly applied, but only transferred to the predicament of the 'concept of information': both the constructivist and the objectivist concept of information form an ordered series, and both series again seem to overlap in human thinking. The constructivist concept of information has pure construction as its basic element and stretches from there to ever more objectified constructions, as in technology. The objectivist concept of information, on the other hand, reaches from the basic elements of objective reality to the most complex phenomena, insofar as a concept of information can be applied to them. In this way, the definitions above the basic level in every case contain specifications of the basic definitions. These two ordered series overlap in human thinking as the activity of construction which, as such, refers to objective reality. If this holds true, a 'science of information' could and should start with an analysis of the concepts of information as found in our human thinking. Here a certain bipolarity can be discovered which is expressed by the transitive and intransitive usages in German: '*informiert werden*' ('being informed') and '*sich informieren*' ('to gain information'). This bipolarity can lay the groundwork for a differentiated concept of information, or, to be more precise, a couple of concepts of information which are linked in their common origin, our active thinking which in its very activity aims at objectivity.

4 THE INFLUENCE OF PHILOSOPHY ON THE UNDERSTANDING OF COMPUTING AND INFORMATION

Klaus Fuchs-Kittowski

Basic Concepts and the Development of Modern Information and Communication Technology

What we consider to be the influence of philosophy on scientific thinking largely depends on how science perceives itself.[1] The understanding and conscious human-oriented design of the relationship between the computer and the creatively active person – i.e. the design of a formal model and the non-formal, natural and social environment – is always more readily recognized as the fundamental philosophical, theoretical and methodological problem of informatics (computer science and information systems).

Informatics/computer science results from the necessity to overcome the tension between technology-based automation, which is based on a purely syntactic interpretation and transformation of information, and creative and active people who carry out semantic information processing based on their knowledge. It is this tension that requires the development and use of user-oriented software and the formal operations to be integrated into complex human work processes. Conceptual strategies that foster the development and integration of modern information technologies into social organization are currently the topic of vivid philosophical and methodological discussions, reflecting the influence of different philosophical schools.

The utilization of information technologies has significantly changed both employee working conditions and the relationship between organizations and their environment. The development of humanity-oriented computer science is a necessary condition for integrating computational systems into social contexts and for largely adapting these systems to the users' needs. The same might also hold true for users who can profit from adapting their behaviours to the conditions of automated processing. Above all, it would be necessary to ensure that information systems will not be explicitly designed according to technical requirements but also take social concepts, values and objectives into account. The question of the relationship between computers and society or between computers and human choices raises the following issues:

1. Differences in the position of human beings compared to the instruments they use;
2. Differences in the baseline with respect to point 1, concerning the automation of intellectual activities;
3. Varying possible directions of the use of computers in different spheres of social life.

In terms of the different positions, we have to take the following ideas into account:

The idea of directly and indirectly regarding human beings and computers as intellectually identical; we can regard this as the position of the researchers following the strand of strong artificial intelligence (AI);

The idea of a mystical exaggeration of human beings' abilities: this is often related to an unjustified criticism of technology (technology pessimism; romantic cultural scepticism);

The idea of combining human beings and machines purposefully and effectively, uniting the advantages of human and machine-performed information processing to form an efficient overall system and giving full consideration to humans beings' creative abilities.

Obviously, the first two ideas represent extreme viewpoints resulting from a unilateral understanding of human beings and computers. We can only overcome this with a dialectical approach. This means that we must determine whether we understand the human being as a disturbing factor, as a relatively imperfect being in comparison to a machine (leading more or less to the idea of replacing human beings with machines) or whether we regard human beings as the genuine masters of these modern technologies. The answer to this question determines the strategies regarding the application of modern information and communication technologies (ICTs) as well as research and training programmes in informatics. Moreover, the answer to this question also influences the development of local and global digital networks such as the internet.

Regarding the Influence of Philosophy on the Guidelines for the Development and Implementation of ICT in Organization

Through our attitude towards information and computers, philosophy has a particular influence on the development, introduction and use of modern ICTs (mediated by the described paradigms) as they occurred at different points in time.[2]

Orientation of Information-System Design and Software Development toward Human Beings

Software has a double (perhaps even a triple) character. It can be understood as a model of an object but also as an informational tool. It is also a medium as a means of communication. At least these three dimensions of software development – object relation, task relation and communication relation – have to be taken into consideration. In the field of software engineering and the modelling connected to it, a decisive paradigm change occurred. It can be characterized as change from product to process orientation in software development.[3]

Here, software development is seen as a specific construction of social reality whose objective is to achieve new levels of intellectual activity by means of computers with an impact on the development of personality and productivity.[4] We can regard this as a reorientation in software engineering from a purely technical perspective to one that encompasses the human being. Obviously, such reorientation can only be achieved by a major shift in the epistemological/methodological basis of this technical domain, a shift that would not have occurred without thinkers who have profoundly reflected on the essence of science and technology.

The guidelines of information-system design have been decisively influenced by the first and second orders of cybernetics. In the 'Declaration of the American Society for Cybernetics', we read: 'Two major orientations have lived side by side in cybernetics from the beginning'. The first orientation arose from the planning of technical systems, which, at the same time, provided models for intelligent processes. AI research developed from this area. The other orientation has focused on the general human question concerning knowledge and, placing it within the conceptual framework of self-organization, has produced, on the one hand, a comprehensive biology of cognition in living organisms (Maturana and Varela) and, on the other, a theory of knowledge construction.[5]

These two orientations have essentially determined the guidelines of ICT development and use. This change regarding the development and application of ICT has been continuously influenced by many differing philosophical positions. The intellectual basis of the first orientation includes:

- Materialistic rationalism (K. Popper)
- Computer theory of mind (J. Fodor and others)
- Black-box functionalism (A. Newell)
- Cybernetics (N. Wiener)
- Formal logic (B. Russel, L. Wittgenstein and others)
- Materialistic and dialectical interpretation of cybernetics (G. Klaus and his students).

In particular, Heinz Zemanek[6] drew our attention to the fact that Ludwig Wittgenstein had anticipated the 'computer with his "Tractatus Logico Philosophicus"'. Accordingly, we should consider the fact that the 'later' Wittgenstein criticized his earlier works as an indicator of the fact that modern ICTs must be based on a broader philosophical basis. The guidelines of development and use of the modern ICTs have changed under the influence of the second orientation. This led to a change from the information-processing paradigm to the paradigm of self-organization and information generation.

The intellectual environment for the new guidelines includes:

- The humanistic orientation of computer science (J. Weizenbaum)
- The reception of the hermeneutics of Heidegger (H. Dreyfus)
- Speech act theory (J. Searle)
- The deeper understanding of Gottlob Frege and Ludwig Wittgenstein (H. Sluga)
- The criticism of AI and the new methodology of system design (T. Winograd and F. Flores)
- The concept of second-order cybernetics (H. von Förster)
- Evolutionary epistemology (J. Piaget, R. Riedel and others)
- Radical constructivism (E. von Glasersfeld)
- The theory of the self-organization of cognition (H. Maturana and F. J. Varela)
- Theoretical neo-structuralism (W. Stegmüller)
- Culturalism (P. Janich)
- The dialectical conception of determinism, evolution theory and activity theory (E. Bloch, G. Klaus, H. Ley and their students)
- The theory of communicative activity (J. Habermas)
- The activity concept of the Russian cultural school as a serious international orientation for computer science/informatics, organizational developers and occupational scientists.
- Constructive realism in informatics and other areas of scholarship (F. Wallner, M. Seel and many scientists)

The conference on *Software Development and Reality Construction*[7] was certainly a milestone in terms of the influence of philosophical thinking on ICT. To my knowledge, it was the first time that computer scientists discussed the philosophical foundation of their proper discipline. Participants started an intensive discussion on the necessity of providing deeper philosophical, theoretical and methodological foundations for ICT, concerning the new philosophical foundation that referred to the tradition of hermeneutics and phenomenology as suggested by Hubert L. Dreyfus[8] and in the work of T. Winograd and F. Flores.[9] These works, in turn, were based on the findings of hermeneutics (Heidegger), speech act theory (Searle), the theory of autopoiese (Matuaran, Varela), and the general discussion of the assumptions of AI by Dreyfus. They also discussed the possible impact of hermeneutics, the theory of self-organization, and activity theory (psychology) on the philosophical foundations of software development and information-systems design. The cultural-historical tradition in activity theory (Vygotsky, Leontjev, Engeström, Hacker, Raethel, Wehner and others)[10] holds special importance for the computer support of knowledge-intensive work processes and scientific work today. On this basis, we can understand that knowledge is not only stored in books or databanks, but also bequeathed by the existing division of labour, the organization of work, and the tools used in the work process (thus, also by the software tools), as well as organizational culture. Accordingly, knowledge is understood as a social process.

Obviously, these contributions and their approaches were too different to establish an agreed view. There was, however, a general understanding of the necessity to overcome the implicit predominant positivistic thinking in ICT. Information-systems design emerged from the conceptual background of the 'rationalistic tradition' in philosophy. A unilateral rationalistic tradition tends to inhibit a deeper understanding of human activities and of the essence of information and organization, of language, decision making, and problem solving, etc. It suggested that it was necessary to overcome the 'naive realism' that originated from a strong 'rational tradition'.

The International Struggle for a Paradigm Change on Philosophical, Theoretical and Practical Bases

Blindly accepting the world of the technical automaton as a model for humankind and corporate organization – in other words, as a model for all reality – has proved to be a dangerous mistake in practice. Philosophical criticism provides the means to realize this mistake.

Linking Semantic and Syntactic Information Processing – Information Centres/Centres of Thought

In the late 1960s, US scientists created the ARPANET, which connected military and academic computers. In the early 1980s, research institutions began to increasingly use this network. The US government transferred the network's operation to the National Science Foundation, which also allowed other countries to gradually connect to the network. This has become a reality since 'the ARPANET is fundamentally connected to and born out of, computer science rather than of the military', as Licklider stated.[11]

At this time we participated in the conference on data communication held by the International Institute for Applied Systems Analysis, where we presented a paper on 'Man/Computer Machine Communication – a Problem of Linking Semantic and Syntactic Information Processing'.[12]

Its main thesis was that problem-solving processes cannot be generally automated. Consequently we need various dialogue forms (direct dialogue and indirect dialogue via an information centre) to support the interaction between human beings and computers. This was based on the insight that man–computer communication is based on a meaningful combination of syntactic (machine) and semantic (human) information processing.

Here, semantic information processing is seen as the combination of information to form new meanings. Typically human intellectual information processing concentrates on the content. Human beings carry out the structural processes, which underlie the meaning of words and sentences – unconsciously.

Syntactic information processing is the structural transformation of information carriers. On the basis of specific rules between information carriers and their meanings, new meanings are ascribed to them. The contents of semantic statements are processed by the mediation of structural transformations.

Some information and computer scientists thought that our strategy clearly showed that the achievements of AI research had been underestimated. To our satisfaction, we experienced exactly the opposite reaction at the IIASA conference, one of the first public meetings for the development of the ARPANET, where our ideas were very well received. This was due to the fact (as we know today) that the development of networks, as conceived by J. C. R. Licklider,[13] leveraged the idea of human beings and machines sharing common features but also manifesting significant differences. It suggested a sensible man–computer interaction (symbiosis). As a prerequisite for this, Licklider saw what he called the 'thinking center'.[14] The leading vision for research on the development of modern digital networks was a technical network that allowed people to cooperate internationally.

Without realizing all the consequences, we were suddenly in the middle of an international dispute that leads *from an understanding of the computer as competitor of human beings to an understanding of the computer as an effective player in a human–computer combination*.[15] However, it was not easy to overcome the original position and to introduce the concept of a purposeful and effective human–computer interaction. During the course of this dispute, J. C. R. Licklider had to fight for his idea of a 'man–computer symbiosis', put forward in relation to the development of new forms of communication to support international collective research. H. Dreyfus wrote in his book, 'What Computers Can't Do; The Limits of Artificial Intelligence', that J. C. R. Licklider favoured the view that 'an interplay of man and machine is presumably most successful'.[16]

Generally, we have to regard the distinction between semantic (associated with human thinking) and automated (syntactic) information processing;

human–machine interaction is a combination of both. It is possible to support this interaction by means of ontologies. A more precise distinction between data, information and knowledge makes clear that it is possible to support this process with modern ICTs. However, this requires a common framework of meaning.

The notion of ontology is determined by logical thinking; contextual relations are excluded. It is likely that formal semantics are also important for people if they apply semantic web applications, for instance, in research on knowledge objects or in specifying search criteria. However, if ontologies are used in the working process – as the necessary communication processes – they have to be interpreted in the framework of existing knowledge, cultural facts and particular activities. To be able to use ontologies for knowledge explication in knowledge-intensive working processes, the ontologies have to relate to a framework of meaning that is common to developers and users, as Christiane Floyd and Stefan Ukena write.[17] In information systems, ontologies are models of partial areas of reality for distinction and new use of knowledge. However, it should be remarked that in addition to formal ontology development, based on strict theories of logic and semantics, there is also a semi-formal approach.

Current developments in the field of ontology engineering can be traced back to early investigations in AI. Because of massive criticism by researchers in this very field[18] as well as by philosophers,[19] computer scientists mostly gave up the research objective to develop a thinking machine, which we usually call 'hard AI'.

Ontology engineers considered themselves to be advanced knowledge engineers, in particular when (in the 1980s) expert systems gained increasing importance. The idea was to overcome the limitations of knowledge engineering, elaborated in intense and heated discussions.[20] Thomas R. Gruber defined ontologies as 'explicit formal specifications of a common conceptualization'.[21] These were supplemented by inference and integrity rules. At that time, the focus of interest was placed on stand-alone systems that were used for problem solving on the base of if–then relationships. However, due to technological change – the development of local and global networks as well as the internet – completely new possibilities emerged, which were accompanied by new questions and problems. At the beginning of the 1980s, AI researchers discussed the question of how to use the knowledge bases for these purposes as well as for ICT-supported cooperative knowledge generation; they connected this with the problems of knowledge reuse and knowledge sharing, or distributed problem solving. It was precisely this type of application that the knowledge bases of the expert systems were not able to provide. The development of ontologies, which has made use of a formal linguistic approach, can also be used for the explication of background knowledge. At the end of the day, a completely new research field emerged from the framework of AI, namely research in ontologies. This new focus describes a shift from knowledge engineering to ontology engineering or from AI research to IA (intelligence amplification or intelligence assistance) research.[22]

Facing the growing demands with respect to ontologies, but also realizing mistakes such as the development of informational ontologies, it gradually becomes clear that we have to render a reference to reality. In order to use ontologies to connect various information systems and to ensure that they are adequately comprehensive and generally accepted by the community of users, a reference to reality appears to be mandatory. Therefore, realism is gaining increasing importance in computer science.[23] Thus, the philosopher Barry Smith rightly demands that computer science should be more clearly oriented toward philosophical investigation, in particular analytical philosophy.[24] Based on his experience in ontology engineering, he explains that bringing data files and the establishment of meta-ontologies (or 'upper ontologies') together will hardly be possible.[25] In this respect activity ontologies might provide a good basis for the establishment of meta-ontologies.[26] Speaking about the influence of philosophy on the understanding of computing and information, here we can see the influence of Gottlob Frege[27] and also of Franz Brentano.

On the Essence and the Evolutionary-State Conception of Information

The theory of informatics/computer science is, and continues to be, characterized by a fundamental tension between formal models and non-formal reality. This theory has to come to terms with the difference of a programme-based structure and the actual dynamics and variety of natural, social and societal life; it must bridge the gap between them by meaningfully embedding automated systems. To achieve this, a thorough discussion of fundamental categories of informatics/computer science is necessary, such as information and organization, storage and memory, and information processing and the genesis of information.

The Variety of Approaches to the Phenomenon and the Essence of Information

Phenomena such as order, information and organization, communicative interaction and directiveness, etc. have not been the subject of traditional natural sciences. In his famous book, *Kybernetik* (Cybernetics), Norbert Wiener described the problematic relation of information physics and its structures when he wrote: 'Information is information, neither matter nor energy. No materialism which does not take this into account can survive the present day.'[28] Here, the idea comes to the fore that with information an effect emerges that goes beyond what has been known to physics thus far. From the quote above, some authors inferred that information is a magnitude that had only been discovered recently and that it was independent of substance and energy. Actually, we already know information as a measurable value, the transformation of which can be described by a formula that is analogous to those we find in physics.[29] Consequently, Szillard, Brillouin and Wiener discussed the connection between physical entropy and information in terms of probability.[30] The similarity between information and entropy, which is expressed in the formula published by Shannon and Weaver in 1949, shows the relationships of information to physics. This does not mean that a better understanding of the role of information in physics is no longer necessary. This range of problems is closely related to the intensively discussed problems of time and the relationship of physics and biology.[31]

In his book, *The Physical Foundation of Biology*,[32] Walter Elsasser called attention to the fact that mechanistic thinking, which claims that all processes obey the laws of physics that can be completely objectified, formalized and – if the latter condition is fulfilled – also programmed, must always presuppose information as a structure that is already given.

We need an understanding of information that does not see information as a given structure. Information science, as well as computer science, AI research and the sciences of cognition should rather take a viewpoint that relates information to the cognitive activity of living organisms.

Today, it is becoming increasingly clear that many of the described ideas depend on each other, as they have been developed in an attempt to understand the phenomenon of information:

The structural understanding of information, in particular that developed by Shannon, Weaver and N. Wiener;

The functional understanding of information, taking into account the receiver's activity, e.g. in E. von Weizsäcker's concept of 'novelty and conformation';

The evolutionary understanding of information as originally suggested by W. Elsasser, M. Eigen and E. Jantsch, as well as by F. J. Varela in his last publication.[33] See also W. Ebeling, K. Fuchs-Kittowski (1998),[34] P. Fleißner, W. Hofkirchner (1997),[35] F. Schweizer and others. K. Haefner (1992)[36] showed the evolution of information-processing systems.

In terms of the influence of philosophy on the understanding of information, we have to consider the influence of Charles W. Morris[37] and Charles S. Peirce[38] for a semiotic understanding of information.

In particular, the evolutionary concept of information is aimed at a new understanding of information through a theory of evolutionary stages. Information has an origin: we do not receive information immediately from the outer world that already exists. Information is relational: information appears as a triple of form (syntax), content (semantic) and effect (pragmatic) generated and used in a multistage process of (in)forming, meaning and evaluation. Information is neither matter nor mind alone, but a link between matter and mind.[39]

Philosophical and Methodological Guidance through the Evolutionary Understanding of Information

The principle of the generation of information has been of fundamental importance for the building of models and theories at the transition zone of physics, chemistry and biology. It must be pointed out that this is also true for building models and theories at the transition zone of computer (software) and the human mind, as well as computer-supported information systems and social organization as a whole.

Many scientists, including Schrödinger and Delbrück, were fascinated by Niels Bohr's lecture 'Light and Life', and conjectured that for the ultimate understanding of life, some novel, fundamental property of matter must first be found, most likely via the discovery of an intuitively paradoxical biological phenomenon. The development of molecular biology showed that such a paradox does not exist. Manfred Eigen[40] clearly said that we do not need a 'new physics' but something 'new in physics' – that is 'information'. We have an information theory but not a theory of 'information generation'.

Most contemporary philosophers of science are familiar with Niels Bohr's 'Copenhagen spirit'. They know the role it has played in the development of modern physics, but only a few of them have taken it seriously as a general perspective towards the world or have recognized its further consequences: the role that the concept of information generation plays in the theory of the origin of life as well as in model and theory formation at the boundaries between physics, chemistry and biology, and now between computer science and the humanities.[41]

The epistemological and methodological implications of the concept of creativity, of information generation, can inspire ideas in nearly all areas of human interest. It provides methodological guidance to navigate between the Scylla of crude reductionism, inspired by nineteenth-century physics and twentieth-century 'mind–brain identity' of (neurophilosophy), strong connectionistic AI research, and the Charybdis of obscurantist vitalism, inspired by nineteenth-century romanticism and twentieth-century functionalistic mind-and-matter/hardware-and-software dualism, of strong cognitivist AI research.

The principle of information generation is also fundamentally important for building models and theories at the transition zone of computer-supported information systems and social organization.

Because of the increasing possibilities regarding the development of the individuality and creativity of people inside and outside the organization, as well as in regard to the expanded possibilities for the transmission of social tradition (through objectification and reification of the knowledge in work organization, tools, software and parts of business culture), enterprises, too, are now on the way to becoming creative learning organizations. In terms of intense interrelation with their environment and the use of internal and external information sources, this means that they are increasingly able to create new information and values in the developing (self-organizing) social organization and to develop new knowledge for new actions internally.[42] The exploitation of these organizational potentials requires an active design of new forms of organization. It means a fundamental move away from the machine/computer models of organization and toward a learning organization. The information-processing approach has led to an understanding of organizations as information-processing systems.[43] The concept of self-organization is bound to the creation of new information, which goes beyond instructive learning. In this respect self-organization appears as the central theoretical concept.[44]

Concrete Humanism as the Basis for Informatics as a Socially Oriented Science

The proper guideline (paradigm) is of fundamental importance because we cannot change or design a world without a clear perspective, without the grasped horizon from which we see it.

Automation as an essential element of scientific and technological progress in our times is definitely not a purely technological problem that can be solved by knowing and mastering the conditions of technological applications. This general insight has led to the development of specific scientific disciplines, such as information and computer science/informatics, including its socially oriented branches.[45] Above all, it is necessary to ensure that the design of automated information systems for economics, education, legal issues and health services are not only derived from technological principles, but that they follow social ideas of values and goals[46] based on a concrete humanism.[47] To develop an information society for all is a very important technical and social goal. Together with the concept of sustainable development, this can be seen as a substantial social innovation.[48] However, this means that an additional task actually lies ahead of us: the task to integrate ICTs in the processes of social and individual development, based on scientifically proven social concepts and humanistic social visions. An example is the critical noosphere vision of Pierre Teilhard de Chardin[49] and Vladimir I. Vernadsky.[50] It is based on the idea that scientific intellect and people's labour transform the existing biosphere into the noosphere.[51] And the idea of noosphere cannot simply be replaced by the terms sociosphere and infosphere since the noosphere expresses a much deeper conception of social development. Although the noosphere is the sphere of human mind and work, it should not be confused with modern technological trends as the technosphere includes the infosphere. However, we can say that the internet and other networks are parts of the growing noosphere.

The vision of the 'noosphere' of Pierre Teilhard de Chardin and V. I. Verna-dsky must instead be seen from the perspective of information and worldwide communication.[52] An open future means that there is no convergent evolution. We have to acknowledge the fact that a simple steering approach using these visions as normative orientations is doomed to fail since no real development process can actually be controlled. The development of a noosphere is not to be seen as the result of a social automatism or teleological process but as a way to grasp the horizon of the possibilities in the development of society. Human beings are and will remain the starting point and the aim of shaping systems of social organization. Starting from the process character of development (Hegel), the inner contradiction of matter as preconditions of the production of new possibilities as an 'ontology of the still not',[53] we have to grasp the horizon of the possibilities for an emancipating design and use of modern ICTs that is oriented toward concrete humanism, so that we can take responsibility for our future.

5 THE EMERGENCE OF SELF-CONSCIOUS SYSTEMS: FROM SYMBOLIC AI TO EMBODIED ROBOTICS

Klaus Mainzer

Classical AI: Symbolic Representation and Control

Knowledge representation, which is today used in database applications, artificial intelligence (AI), software engineering and many other disciplines of computer science has deep roots in logic and philosophy.[1] In the beginning, there was Aristotle (384 BC–322 BC) who developed logic as a precise method for reasoning about knowledge. Syllogisms were introduced as formal patterns for representing special figures of logical deductions. According to Aristotle, the subject of ontology is the study of categories of things that exist or may exist in some domain.

In modern times, Descartes considered the human brain as a store of knowledge representation. Recognition was made possible by an isomorphic correspondence between internal geometrical representations (*ideae*) and external situations and events. Leibniz was deeply influenced by these traditions. In his *mathesis universalis*, he required a universal formal language (*lingua universalis*) to represent human thinking by calculation procedures and to implement them by means of mechanical calculating machines. An *ars iudicandi* should allow every problem to be decided by an algorithm after representation in numeric symbols. An *ars iveniendi* should enable users to seek and enumerate desired data and solutions of problems. In the age of mechanics, knowledge representation was reduced to mechanical calculation procedures.

In the twentieth century, computational cognitivism arose in the wake of Turing's theory of computability. In its functionalism, the hardware of a computer is related to the wetware of the human brain. The mind is understood as the software of a computer. Turing argued that if the human mind is computable, it can be represented by a Turing programme (Church's thesis) which can be computed by a universal Turing machine, i.e. technically by a general purpose computer. Even if people do not believe in Turing's strong AI-thesis, they often claim classical computational cognitivism in the following sense: computational processes operate on symbolic representations referring to situations in the outside world. These formal representations should obey Tarski's correspondence theory of truth. Imagine a real-world situation X1 (e.g. some boxes on a table) which is encoded by a symbolic representation A1 = encode (X1) (e.g. a description of the boxes on the table). If the symbolic representation A1 is decoded, then we get the real-world situation X1 as its meaning, i.e. decode (A1) = X1. A real-world operation T (e.g. a manipulation of the boxes on the table by hand) should produce the same real-world result A2, whether performed in the real world or on the symbolic representation: decode(encode(T)(encode(X1))) = T(X1) = X2. Thus, there is an isomorphism between the outside situation and its formal representation in the Cartesian tradition. As the symbolic operations are completely determined by algorithms, the real-world processes are assumed to be completely controlled. Therefore, classical robotics operate with completely determined control mechanisms.

New AI: Self-Organization and Controlled Emergence

Knowledge representations with ontologies, categories, frames and scripts of expert systems work along this line. However, they are restricted to a specialized knowledge base without the background knowledge of a human expert. Human experts do not rely on explicit (declarative) rule-based representations, but on intuition and implicit (procedural) knowledge.[2] Furthermore, as Wittgenstein knew, our understanding depends on situations. The situatedness of representations is a severe problem of informatics. A robot, for example, needs a complete symbolic representation of a situation which must be updated if the robot's position is changed. Imagine that it surrounds a table with a ball and a cup on it. A formal representation in a computer language may be ON(TABLE,BALL), ON(TABLE,CUP), BEHIND(CUP,BALL), etc. Depending on the robot's position relative to the arrangement, the cup is sometimes behind the ball and sometimes not. So, the formal representation BEHIND(CUP,BALL) must always be updated in changing positions. How can the robot prevent incomplete knowledge? How can it distinguish between reality and its relative perspective? Situated agents like human beings need no symbolic representations and updating. They look, talk and interact bodily, for example by pointing to things. Even rational acting in sudden situations does not depend on internal representations and logical inferences, but on bodily interactions with a situation (for example looking, feeling, reacting).

Thus, we distinguish formal and embodied acting in games with more or less similarity to real life: chess, for example, is a formal game with complete representations, precisely defined states, board positions and formal operations. Soccer is a non-formal game with skills depending on bodily interactions, without complete representations of situations, and operations which are never exactly identical. According to Merleau-Ponty, intentional human skills do not need any internal representation, but they are trained, learnt and embodied in an optimal *gestalt* which cannot be repeated.[3] An athlete like a pole-vaulter cannot repeat her successful jump like a machine generating the same product. Husserl's representational intentionality is replaced by embodied intentionality. The embodied mind is no mystery. Modern biology, neural and cognitive science give many insights into its origin during the evolution of life.

The key concept is self-organization of complex dynamical systems.[4] The emergence of order and structures in nature can be explained by the dynamics and attractors of complex systems. They result from collective patterns of interacting elements in the sense of many-bodies problems that cannot be reduced to the features of single elements in a complex system. Nonlinear interactions in multicomponent ('complex') systems often have synergetic effects, which can neither be traced back to single causes nor be forecasted in the long run or controlled in all details. The whole is more than the sum of its parts. This popular slogan for emergence is precisely correct in the sense of nonlinearity.

The mathematical formalism of complex dynamical systems is taken from statistical mechanics. If the external conditions of a system are changed by varying certain control parameters (for example temperature), the system may undergo a change in its macroscopic global states at some critical point. For instance, water as a complex system of molecules changes spontaneously from a liquid to a frozen state at a critical temperature of 0 Celsius. In physics, those transformations of collective states are called phase transitions. Obviously they describe a change of self-organized behaviour between the interacting elements of a complex system. The suitable macrovariables characterizing the change of global order are denoted as 'order parameters'. They can be determined by a linear-stability analysis.[5] From a methodological point of view, the introduction of order parameters for modelling self-organization and the emergence of new structures is a giant reduction of complexity. The study of, perhaps, billions of equations, characterizing the behaviour of the elements on the microlevel, is replaced by some few equations of order parameters, characterizing the macrodynamics of the whole system. Complex dynamical systems and their phase transitions deliver a successful formalism to model self-organization and emergence. The formalism does not depend on special, for example, physical laws, but must be appropriately interpreted for different applications.

There is a precise relation between self-organization of nonlinear systems with continuous dynamics and discrete cellular automata. The dynamics of nonlinear systems is given by differential equations with continuous variables and a continuous parameter of time. Sometimes, difference equations with discrete time points are sufficient. If even the continuous variables are replaced by discrete (e.g. binary) variables, we get functional schemes of automata with functional arguments as inputs and functional values as outputs. There are classes of cellular automata modelling attractor behaviour of nonlinear complex systems which is well-known from self-organizing processes. But in many cases, there is no finite programme, in order to forecast the development of random patterns. Thus, pattern emergence cannot be controlled in any case. Self-organization and pattern emergence can also be observed in neural networks, working like brains with appropriate topologies and learning algorithms. A simple robot with diverse sensors (e.g. proximity, light, collision) and motor equipment can generate complex behaviour by a self-organizing neural network. In the case of a collision with an obstacle, the synaptic connections between the active nodes for proximity and collision layer are reinforced by Hebbian learning: A behavioural pattern emerges, in order to avoid collisions in future.

Obviously, self-organization leads to the emergence of new phenomena on sequential levels of evolution. Nature has demonstrated that self-organization is necessary, in order to manage the increasing complexity on these evolutionary levels. But nonlinear dynamics can also generate chaotic behaviour which cannot be predicted and controlled in the long run. In complex dynamical systems of organisms, monitoring and controlling are realized on hierarchical levels. Thus, we must study the nonlinear dynamics of these systems in experimental situations, in order to find appropriate order parameters and to prevent undesired emergent behaviour as possible attractors. The challenge of complex dynamical systems is controlled emergence.

A key application is the nonlinear dynamics of brains. Brains are neural systems which allow quick adaption to changing situations during the lifetime of an organism. In short: they can learn. The human brain is a complex system of neurons self-organizing in macroscopic patterns by neurochemical interactions. Perceptions, emotions, thoughts and consciousness correspond to these neural patterns. Motor knowledge, for example, is learnt in an unknown environment and stored implicitly in the distribution of synaptic weights of the neural nets. In the human organism, walking is a complex bodily self-organization, largely without central control of brain and consciousness: it is driven by the dynamical pattern of a steady periodic motion, the attractor of the motor system. Motor intelligence emerges without internal symbolic representations.

What can we learn from nature? In unknown environments, a better strategy is to define a low-level ontology, introduce redundancy – and there is a lot in the sensory systems, for example – and leave room for self-organization. Low-level ontologies of robots only specify systems like the body, sensory systems, motor systems and the interactions among their components, which may be mechanical, electrical, electromagnetic, thermal, etc. According to the complex systems approach, the components are characterized by certain microstates generating the macrodynamics of the whole system.

Take a legged robot. Its legs have joints that can assume different angles, and various forces can be applied to them. Depending on the angles and the forces, the robot will be in different positions and behave in different ways. Further, the legs have connections to one another and to other elements. If a six-legged robot lifts one of the legs, this changes the forces on all the other legs instantaneously, even though no explicit connection needs to be specified.[6] The connections are implicit: they are enforced through the environment, because of the robot's weight, the stiffness of its body, and the surface on which it stands. Although these connections are elementary, they are not explicit and are only included if the designer wishes to include them. Connections may exist between elementary components that we do not even realize. Electronic components may interact via electromagnetic fields that the designer is not aware of. These connections may generate adaptive patterns of behaviour with high fitness degrees (order parameter). But they can also lead to sudden instability and chaotic behaviour. In our example, communication between the legs of a robot can be implicit. In general, much more is implicit in a low-level specification than in a high-level ontology. In restricted simulated agents with bounded knowledge representation, only what is made explicit exists, whereas in the complex real world, many forces exist and properties obtain, even if the designer does not explicitly represent them. Thus, we must study the nonlinear dynamics of these systems in experimental situations, in order to find appropriate order parameters and to prevent undesired emergent behaviour as possible attractors.

It is not only 'low-level' motor intelligence, but also 'high-level' cognition (e.g. categorization) that can emerge from complex bodily interaction with an environment by sensory-motor coordination without internal symbolic representation. We call it 'embodied cognition': an infant learns to categorize objects and to build up concepts by touching, grasping, manipulating, feeling, tasting, hearing and looking at things, and not by explicit representations. The categories are based on fuzzy patchworks of prototypes and may be improved and changed during life. We have an innate disposition to construct and apply conceptual schemes and tools (in the sense of Kant).

Moreover, cognitive states of persons depend on emotions. We recognize emotional expressions of human faces with pattern recognition of neural networks and react by generating appropriate facial expressions for non-verbal communication. Emotional states are generated in the limbic system of the brain which is connected with all sensory and motoric systems of the organism. All intentional actions start with an unconscious impulse in the limbic system which can be measured some fractals of a second before their performance. Thus, embodied intentionality is a measurable feature of the brain.[7] Humans use feelings to help them navigate the ontological trees of their concepts and prefer-ences, to make decisions in the face of increasing combinatorical complexity: emotions help to reduce complexity.

The embodied mind[8] is obviously a complex dynamical system acting and reacting in dynamically changing situations. The emergence of cognitive and emotional states is made possible by brain dynamics which can be modelled by neural networks. According to the principle of computational equivalence,[9] any dynamical system can be simulated by an appropriate computational sys-tem. But, contrary to Turing's AI thesis, that does not mean computability in any case. In complex dynamical systems, the rules of locally interacting elements (e.g. Hebb's rules of synaptic interaction) may be simple and programmed in a computer model. But their nonlinear dynamics can generate complex patterns and system states which cannot be forecast in the long run without increasing loss of computability and information. Thus, artificial minds could have their own intentionality, cognitive and emotional states which cannot be forecast and computed as in the case of natural minds. Limitations of computability are char-acteristic features of complex systems.

In a dramatic step, the complex systems approach has been enlarged from neural networks to global computer networks like the World Wide Web. The internet can be considered as a complex open computer network of autono-mous nodes (hosts, routers, gateways, etc.), self-organizing without central mechanisms. Routers are nodes of the network determining the local path of each information packet by using local routing tables with cost metrics for neighbouring routers. These buffering and resending activities of routers can cause congestion on the internet. Congested buffers behave in surprising anal-ogy to infected people. There are nonlinear mathematical models describing true epidemic processes like malaria extension as well as the dynamics of routers. Computer networks are computational ecologies.[10]

But complexity of global networking does not only mean increasing numbers of PCs, workstations, servers and supercomputers interacting via data traffic on the internet. Below the complexity of a PC, low-power, cheap and smart devices are distributed in the intelligent environments of our everyday world. Like GPS in car traffic, things in everyday life could interact telematically by sensors. The real power of the concept does not come from any one of these single devices. In the sense of complex systems, the power emerges from the collective interaction of all of them. For instance, the optimal use of energy could be considered as a macroscopic order parameter of a household realized by the self-organizing use of different household goods according to less consumption of electricity during special time periods with cheap prices. The processors, chips and displays of these smart devices don't need a user interface like a mouse, windows or keyboards, but just a pleasant and effective place to get things done. Wireless computing devices on small scales become more and more invisible to the user. Ubiquitous computing enables people to live, work, use and enjoy things directly without being aware of their computing devices.

Self-Conscious Systems, Human Responsibility and Freedom

Obviously, interacting embodied minds and embodied robots generate embodied superorganisms of self-organizing information and communication systems. What are the implications of self-organizing human–robot interaction (HRI)? Self-organization means more freedom, but also more responsibility. Controlled emergence must be guaranteed in order to prevent undesired side effects. But, in a complex dynamical world, decision making and acting is only possible under conditions of bounded rationality. Bounded rationality results from limitations on our knowledge, cognitive capabilities and time. Our perceptions are selective, our knowledge of the real world is incomplete, our mental models are simplified, our powers of deduction and inference are weak and fallible. Emotional and subconscious factors affect our behaviour. Deliberation takes time and we must often make decisions before we are ready. Thus, knowledge representation must not be restricted to explicit declarations. Tacit background knowledge, change of emotional states, personal attitudes and situations with increasing complexity are challenges of modelling information and communication systems. Human-oriented information services must be improved in order to support a sustainable information world.

While the computational process of a PC, for example, is running, we often know neither the quality of the processing nor how close the current processing is to a desired objective. Computational processes seldom have intermediate results to tell us how near the current process is to any desired behaviour. In biological systems, for example, we humans experience a sense that we know the answer. In a kind of recursive self-monitoring, some internal processes observe something about our cognitive states that help us to evaluate our progress. The evolutionary selection value of self-reflection is obvious: if we have these types of observations available to us, we can alter our current strategies according to changing goals and situations. Engineered systems have some counterparts to the kinesthetic feedback one finds in biological systems. But the challenge is to create feedback that is meaningful for decisions in a system that can reflectively reason about its own computations, resource use, goals and behaviour within its environment. This kind of cognitive instrumentation of engineered systems[11] can only be the result of an artificial evolution, because cognitive processes of humans with their multiple feedback processes could also only develop during a long history of evolution and individual learning.

Thus, we need generative processes, cognitive instrumentation and reflective processes of systems in order to handle the complexity of human–robot interactions. Biological systems take advantage of layers with recursive processing of self-monitoring and self-controlling from the molecular and cellular to the organic levels. Self-reflection leads to a knowledge that is used by a system to control its own processes and behaviours. What distinguished self-reflection from any executive control process is that this reflection involves reasoning about that system, being able to determine or adjust its own goals because of this reflection. But self-reflection must be distinguished from self-consciousness. Consciousness is at least partly about the feeling and experience of being aware of one's own self. Therefore, we could construct self-reflecting systems without consciousness that may be better than biological systems for certain applications. It is well known that technical instruments (e.g. sensors) already surpass the corresponding capacities of natural organisms with many orders of magnitude. Self-reflecting systems could help to improve self-organization and controlled emergence in a complex world.

But, how far should we go? Self-consciousness and feeling are states of brain dynamics which could, at least in principle, be simulated by computational systems. The brain does not only observe, map and monitor the external world, but also internal states of the organism, especially its emotional states. Feeling means self-awareness of one's emotional states. In neuromedicine, the 'Theory of Mind' (ToM) even analyses the neural correlates of social feeling which are situated in special areas of the neocortex. People, for example those suffering from Alzheimer's disease, lose their feeling of empathy and social responsibility because the correlated neural areas are destroyed. Therefore, our moral reasoning and decision making have a clear basis in brain dynamics which, in principle, could be simulated by self-conscious artificial systems. In highly industrialized nations with advanced aging, feeling robots with empathy may be a future perspective for nursing old people when the number of young people engaged in public social welfare has decreased and the personal costs have increased dramatically.

Humans are not at the centre of the universe and evolution, but they are at the centre of their history and culture. The concept of human personality refers to human historicity, self-identity, intentionality and embedding in the intimacy of human social and cultural identity. Therefore, AI, cognitive science and computer science have to take care of humans as a value and purpose on its own (Kant: 'self-purpose') which should be the measure of human technology. That is a postulate of practical reason which has developed and approved itself in evolution and the history of mankind. In principle, future technical evolution could generate self-conscious systems which are not only human copies ('clones'), but artificial organisms with their own identity and intimacy which would differ from ours. But why should we initiate an evolution separate from human interests? AI, biotechnology, information technology and communication technology should be developed as a human service in the tradition of medicine in order to heal and to help.

This is a humanistic vision different from science fiction dreams which only trust in the technical dynamics of increasing computational capacity, leading automatically to eternal happiness and computational immortality. In nonlinear dynamics, there is no guarantee of final stable states of order. We need 'order parameters' for moral orientation in changing situations of our development. We should not trust in the proper dynamics of evolution, and we should not accept our deficient nature which is more or less the result of a random biological game and compromise under changing conditions on the earth. The dignity of humans demands to interfere, change and improve their future. But, it should be our decision who we want to be in future, and which kind of artificial intelligence and artificial life we need and want to accept beside us.

6 ARTIFICIAL INTELLIGENCE AS A NEW METAPHYSICAL PROJECT

Aziz F. Zambak

Introduction

It has been a long-standing philosophical issue whether machines can think or not. In the history of philosophy, this issue is discussed using different types of questions such as 'Is it possible to design intelligent artefacts?', 'Can a mechanistic performance of a machine imitate human intelligence?' or 'Can reasoning be reduced to some kind of calculation?'. The philosophy of artificial intelligence can be considered a modern aspect of these discussions. Artificial intelligence (hereafter AI) is a field of interdisciplinary study that lies at the intersection of cognitive science, linguistics, logic, neuroscience, computer science and psychology.

The definition and the goal of AI were specified in the middle of the twentieth century, and the study of AI has developed quickly in a very short period of time. These developments can be observed in certain branches of industry, medicine, education, the military, communication, game playing and translating. Such fields of application of AI comprise its technological aspect. AI also has a scientific aspect that deals with theoretical, methodological and conceptual questions, and the philosophy of AI deals with the latter.

The Definition and Aim of AI

What is AI? This is not an easy question, and it is not possible to find an exact definition accepted by all AI researchers. AI has various definitions. This variety is caused by divergent views with regard to the aim and scope of AI. There are four different approaches that cause different understandings of the aim and scope of AI:

The technological approach: AI researchers design computer systems that are already in commercial use. In the technological approach, AI is the name of a single and specific technical (machine) project that aims to produce a specific product. Therefore, AI is more like engineering (applied techniques) than cognitive theory (pure science). Generally, the applied techniques in AI deal with problems dominated by combinatorial explosion.[1] AI is concerned with designing computer systems to succeed at certain practical and intelligent tasks in the real world. Since the technological approach to AI focuses on the heuristic character of human intelligence, AI researchers study problem-solving techniques.

The imitation approach: After some technical developments in computer science from the early 1950s to the late 1960s, the early years of AI were a period of enthusiasm and great expectations. In this period, many AI researchers made optimistic predictions for the future of computer science and machine research.[2] These optimistic predictions were very determinative in their definitions of AI. General characteristics of these definitions included the idea that a machine can be made to simulate human intelligence and duplicate psychological phenomena. In these definitions, the notions of simulation and duplication are based on the assumption that every aspect of cognitive processes, skills and other features of the human mind can in principle be precisely comprehended.

The intermediary approach: Other AI researchers put much more stress on the task of constructing models and programmes that are useful tools for the study of human cognitive activity. Therefore, in the intermediary approach, the aim of AI is to understand human intelligence/thinking by using computer (machine) techniques (models); AI is therefore a methodological tool for the very general investigation of the nature of the human mind. The intermediary approach can be considered the main reason behind the current dominance of the computational theories among researchers in cognitive science.

The expert-system approach: In AI, the early 1970s was the period of awakening from a sweet dream. After disappointments in certain projects and harsh criticisms against AI, the tasks in AI were changed. After the 1970s, the general themes were knowledge-based systems, expert systems and connectionist networks. These shifts in tasks and themes within computer science led to a new understanding of AI.

In spite of these divergent approaches and definitions of AI, the basic and common features of it can be reformulated as follows:

- AI is a machine *performance* that can be ascribed to 'intelligence' and 'mental states'.
- AI is interested in human intelligence, behaviour and mental states as a pattern or form in a functional way.
- AI is a useful tool for explaining certain qualities of human cognitive abilities.
- AI aims to imitate the essential faculties and powers of human intelligence.

In our opinion, *AI is the study of constructing machine intelligence in an agentive manner [position]*. There are two essential points in this definition that should be explained: First, man is not a machine and a machine is not a man. Our goal of studying AI is not to imitate (or simulate) the human mind as a machine; but to situate the *conditions*[3] of the human mind into machine intelligence. We prefer to use the term 'machine intelligence'[4] rather than 'artificial intelligence'. Although machine intelligence is the practice of AI systems, it is different from the mainstream, which we have described above, in AI. Intelligence is not a single phenomenon that can be represented in a unified formal system. We defend the idea that machine intelligence does not necessarily process information in the way humans do. In other words, questions like *how does the mind work, how are problems solved, how are decisions made, how are patterns detected, how is memory structured, how is knowledge organized* and *how are statements presented via language* can be studied in a different methodological and theoretical perspective. We see machine intelligence as an alternative methodological and theoretical field for the study of these questions. Machines and humans are not species of the same cognitive structure, but they can be the subject of the same questions about cognition, information, cognitive skills, mental states, and so on. The definition of AI that we have proposed rejects a human-centric understanding of intelligence and cognitive skills. In our definition, the emphasis on machine intelligence mentions that there can be alternative ways to define certain cognitive (mental) concepts, such as creativity, subjectivity, experience, free will, intention, cognition, learning, reasoning, imagining, communication, perception, memory, etc. in an original machine-based model. A machine can have intelligence and other cognitive structures in its peculiar style. Intelligence in machines can be respected and studied for what it is. The notion of 'machine intelligence' does mention that machines and humans may have different ways of thinking, perceiving, memorizing, understanding, learning, using language and playing chess. Second, in our definition, we have mentioned agency which is (or must be) the essential notion in AI. Agency is the basic *condition* that allows machine intelligence and the human mind to be the subjects of the same kinds of questions about intelligence and other cognitive structures.

Philosophy and AI

Philosophy is relevant to AI, and this relevance is constructed from different perspectives. For instance, Ringle defines the relevance of philosophy to AI in two key areas: '(a) the ontological problem of the relationship between minds and bodies; and (b) the epistemological problem of the analysis of mentalistic terms'.[5] Sayre describes the value of AI to philosophy in a methodological sense.[6] Kyburg considers AI as a philosophical laboratory: 'This is a wonderful time to be doing philosophy ... even in such an apparent simple area as the construction of cognitive agents, there is plenty of plain old-fashioned philosophical thinking and analysis to be done.[7] Darden sees AI as an experimental epistemology: 'AI systems allow philosophers to experiment with knowledge and reasoning ... It investigates methods for representing knowledge and for modeling reasoning strategies that can manipulate that knowledge.'[8] McCarthy and Hayes claim that AI includes major traditional and modern specific problems of philosophy.[9] According to Ford et al., there is a very close relation between AI and philosophy: 'AI and philosophy have things to say to one another. Any attempt to create and understand minds must be a philosophical interest. In fact, AI is philosophy, conducted by novel means.'[10] Boden mentions that AI is relevant to philosophy since AI suggests 'solutions to many traditional problems in the philosophy of mind, and for illuminating the hidden complexities of human thinking and personal psychology'.[11]

In modern AI, there are many approaches that consider AI to be a new way of philosophizing. For instance, Ford et al. use the term *android epistemology*[12] in order to describe AI as a novel philosophical methodology. Steinhart describes AI as *digital metaphysics*[13] and argues that computation, a dominant approach in AI, is the *foundation* of metaphysics: '*ultimate reality is a massively parallel computing machine sufficiently universal for the realization of any physically possible world. Ultimate reality is computational space-time.*'[14] Boden considers AI as a proper discipline for conceptualizing the mind.[15] McCorduck claims that AI provides a better explanation for whole behavioural and natural events since AI's conceptual framework, including information, computation, coding, storage, etc., provides a dynamic and open-ended understanding of behaviour and nature.[16] Ringle argues that AI offers alternative strategies for certain epistemological problems.[17] Pollock defends the idea that AI can change epistemology. He makes a distinction between procedural epistemology and descriptive epistemology. He describes procedural epistemology as one of the main characteristics of AI:

> Procedural epistemology is concerned with the procedures comprising rational epistemic cognition, and the main way in which computers impact procedural epistemology is by providing the tool for constructing computer models of proposed theories. Such a model becomes an AI system.[18]

Sloman considers AI to be a leading discipline for a novel understanding of philosophical problems:

> The best way to make substantial new progress with old philosophical problems about mind and body, about perception, knowledge, language, logic, mathematics, science and aesthetics, is to reformulate them in the context of an attempt to explain the possibility of a mind. The best way to do this is to attempt to *design* a working mind, i.e. a

mechanism which can perceive, think, remember, learn, solve problems, interpret symbols or representations, use language, act on the basis of multiple motives, and so on.[19]

Thagard argues that AI introduces new conceptual resources in order to deal with certain problems in epistemology and philosophy of science.[20] Doyle sees AI as the study of all possible minds. He argues that AI is the conceptual and theoretical investigation of psychology by means of computational concepts.[21]

AI as a New Metaphysical Project

It is a general tendency to view AI as a modern aspect of traditional epistemological and ontological problems. In addition to that, philosophers who see a close relation between AI and philosophy are hopeful that AI can make contributions to the solutions and analysis of certain philosophical problems and concepts. I go one step further and claim that AI is an intellectual revolution and can be considered a new metaphysical project. In philosophy, AI can be an *intellectual source* and *experimental field* for developing a novel notion of 'cognitive structure' within which various forms of human knowledge can be presented and altered. AI is the most comprehensive and/or hypothetical study for the possibility of cognition and intelligence. Whereas the *agent-as-we-know-it* is the central epistemological and ontological concern in philosophy, AI's scope grasps all of *agent-as-it-could-be*. Philosophy is concerned with constant forms of agency and cognition, whereas AI is concerned with dynamic and conceivable ones. Therefore, AI gives us new possibilities for thinking in terms of a novel form of agency, cognition, mind, reasoning, creativity, free will, subjectivity, language, etc. Sloman considers AI a new way of thinking about ourselves:

> [AI] can change our thinking about ourselves: giving us new models, metaphors, and other thinking tools to aid our efforts to fathom the mysteries of the human mind – and heart. The new discipline of Artificial Intelligence is the branch of computing most directly concerned with this revolution. By giving us new, deeper, insights into some of our inner processes, it changes our thinking about ourselves. It therefore changes some of our inner processes, and so changes what we are, like all social, technological and intellectual revolutions.[22]

AI not only is a result of a rich intellectual tradition in mathematics, physics, psychology and philosophy but also contributes to them. AI has many novel models and methods to offer philosophers for solving and analysing certain philosophical problems from a different perspective.

The original theoretical models and methods in AI give the possibility of reaching a new metaphysics that deals with certain philosophical problems. Agency is the essential and constitutive notions behind these models and methods. In other words, what makes AI an alternative strategy or a novel conceptual resource for philosophy is the possibility of designing a mind that has the characteristics of agency.

Agency

Agency must be the central notion in AI since the cognition of reality originates from agentive actions. We claim that agency is the ontological and epistemo-logical constituent of reality and cognition. Reality is characterized by agentive activity. In the metaphysics of AI, reality must not be seen as a mere psychic given or a datum of a mental state. On the contrary, it is an embodiment in which the subject and his surrounding environment should be situated in an agentive rela-tion. Therefore, agency is primary, even in defining objectivity. Mental activity and intelligence are not located in the organism; they are not an inner and pri-vate activity of the organism. Intelligence is not a primitive capacity, but is rather something achieved by agentive actions. To become conscious is to be able to act in an agentive manner. Thought culminates in a form of agentive cognition and in AI, agentive cognition is the only genuine form of knowledge.

Agency must be the essential criterion for the success of machine intelligence instead of linguistic-behavioural-based criteria (for instance, the Turing test). Since agency is the system of actions in which mind is rooted, it – in AI – is the basic constituent of rationality, intelligence, mental acts and other cognitive skills. In other words, agentive action is the primary source for the rationali-zation and cognitive processes in machine intelligence. However, in order to prevent misunderstanding, we must mention that every action is not essentially or originally agentive. In AI, action cannot be seen only as a response to external stimuli. Action must be an interactional process that machine intelligence does for a reason.[23] The essence of agentive action is rationalization in which machine intelligence acts in order to achieve its goals.

AI must consider mental activity as a form of action of a dynamic represen-tational system, developed during *interaction* within the environment. Equating properties of the mind with properties of its elements is the basic mistake. Mental activity cannot be a subject of a special localization of the brain or certain prop-erties of neurons. Behaviour and agentive actions cannot be found in locations in the brain, but in the whole agent–environment interaction system. Therefore, in AI, it is possible to replace dynamic representational systems with physiological properties of the brain, producing behavioural and agentive results. The occur-rence of mental activity in machine intelligence does mean a new kind of action of the highly dynamic representational system capable of making inferences from its experiences in order to achieve new results of action and form novel systems directed towards the future. Therefore, in AI, mental activity is not a mystical emergent property of neural elements, but a form of agentive action necessarily following from the development of a dynamic representational system.

Agency consists of inherently relational activities, aimed at exerting a cer-tain influence on the environment. Therefore, in AI, we propose descriptions of 'mental activity' and 'representation' in environment-referential instead of neuron-identified terms. We claim that theoretical studies of cognitive science have consistently been based on the idea that the mind and surroundings form two distinct systems and that mental activity is situated in the mind, that it is an inner and subjective activity of the mind. It is this main presupposition that seems to lead up a blind alley in cognitive science and AI. This presupposition leads cognitive science and AI researchers to the idea that the formation of cognition depends on transmission of information from the surroundings to the mind. However, we defend a different position in which we consider the mind and the surroundings as one system; all formation and increase of cognition means only dynamic reorganization or expansion of this system. In AI, cognition must be *created* in an agentive manner; it cannot be transmitted or moved from one head to another. Instead of focusing on the linear sequence of information, AI research should be directed toward the *conditions* necessary for generating information. Agentive cognition is an ongoing informational process based on a mutual constitutive relationship between the mind and its surroundings. Agentive cognition is not just a formal representation correlated with a sequence of information, but instead refers to certain aspects of a mind-surroundings system as a whole.

7 THE RELEVANCE OF PHILOSOPHICAL ONTOLOGY TO INFORMATION AND COMPUTER SCIENCE

Barry Smith

Historical Background

Ontology as a branch of philosophy is the science of what is, of the kinds and structures of objects, properties, events, processes and relations in every area of reality. The earliest use of the term 'ontology' (or '*ontologia*') seems to have been in 1606 in the book *Ogdoas Scholastica* by the German Protestant scholastic Jacob Lorhard. For Lorhard, as for many subsequent philosophers, 'ontology' is a synonym of 'metaphysics' (a label meaning literally: 'what comes after the *Physics*'), a term used by early students of Aristotle to refer to what Aristotle himself called 'first philosophy'. Some philosophers use 'ontology' and 'metaphysics' to refer to two distinct, though interrelated, disciplines, the former to refer to the study of what *might* exist; the latter to the study of which of the various alternative possible ontologies is in fact true of reality.[1]

The upper level of Lorhard's own ontology is illustrated in Figure 7.1.[2]

The term – and the philosophical discipline of ontology – has enjoyed a chequered history since 1606, with a significant expansion, and consolidation, in recent decades. We shall not discuss here the successive rises and falls in philosophical acceptance of the term, but rather focus on certain phases in the history of recent philosophy which are most relevant to the consideration of its recent advance, and increased acceptance, also outside the discipline of philosophy.

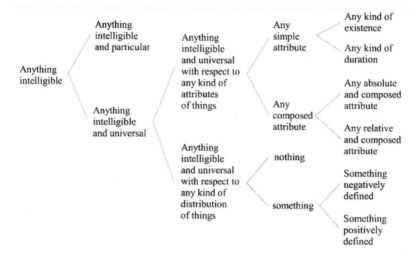

Figure 7.1: Top level of Lorhard's ontology.

Varieties of Philosophical Ontology

For the philosophical ontologist, ontology seeks to provide a definitive and exhaustive classification of entities in all spheres of being. The classification should be definitive in the sense that it can serve as an answer to such questions as: what classes of entities are needed for a complete description and explanation of all the goings-on in the universe? Or: what classes of entities are needed to give an account of what makes true all truths? It should be exhaustive in the sense that all types of entities should be included in the classification, and it should include also all the types of relations by which entities are tied together to form larger wholes.

Different schools of philosophy offer different approaches to the provision of such classifications. One large division is that between what we might call substantialists and fluxists, which is to say between those who conceive ontology as a substance- or thing- (or continuant-) based discipline and those who favour an ontology centred on events or processes (or occurrents). Another large division is between what we might call adequatists and reductionists. Adequatists seek a taxonomy of the entities in reality at all levels of aggregation, from the microphysical to the cosmological, and including also the middle world (the *mesocosmos*) of human-scale entities (carpets, caves, caravans, carpal tunnel syndromes) in between. Reductionists see reality in terms of some privileged level of existents, normally the smallest. They thereby seek to establish the 'ultimate furniture of the universe' by decomposing reality into its simplest constituents, or they seek to 'reduce' in some other way the apparent variety of types of entities existing in reality, often by providing recipes for logically translating assertions putatively about entities at higher levels into assertions allowable from the reductionist perspective.

In the work of adequatist philosophical ontologists such as Aristotle, Ingarden,[3] Johansson,[4] Chisholm[5] and Lowe,[6] the proposed taxonomies are in many ways comparable to those produced and used in empirical sciences such as biology or chemistry, though they are of course radically more general than these. Adequatism – which is the view defended also by the author of this essay – transcends the dichotomy between substantialism and fluxism, since its adherents accept categories of both continuants and occurrents.

Ontology, for the adequatist, is a descriptive enterprise. It is distinguished from the special sciences not only in its radical generality but also in its goal: the ontologist seeks not prediction, but rather description of a sort that is based on adequate classification. Adequatists study the totality of those objects, properties, processes and relations that make up the world on different levels of granularity, whose different parts and moments are studied by the different scientific disciplines – often, as in the case of all the adequatists listed above, with a goal of providing the philosophico-ontological tools for the unification or integration of science.

Methods of Ontology

The methods of ontology in philosophical contexts include the development of theories of wider or narrower scope and the refinement of such theories by measuring them either against difficult counterexamples or against the results of science. These methods were already familiar to Aristotle.

In the course of the twentieth century a range of new formal tools became available to ontologists for the development, expression and refinement of their theories. Ontologists nowadays have a choice of formal frameworks (deriving from algebra, category theory, mereology, set theory, topology) in terms of which their theories can be formulated. These new formal tools, along with the languages of formal logic, allow philosophers to express intuitive principles and definitions in clear and rigorous fashion, and, through the application of the methods of formal semantics, they can allow also for the testing of theories for consistency and completeness. When we examine the work of computational ontologies below, we shall see how they have radicalized this approach, using formal methods as implemented in computers as a principal method of ontology development.

The Role of Quine

Some philosophers have thought that the way to do ontology is exclusively through the investigation of scientific theories. With the work of Quine there arose in this connection a new conception of the proper method of philosophical ontology, according to which the ontologist's task is to establish what kinds of entities scientists are committed to in their theorizing.[7]

Quine took ontology seriously. His aim was to use science for ontological purposes, which means: to find the ontology *in* scientific theories. Ontology is for him a network of claims (a web of beliefs) about what exists, deriving from the natural sciences. Each natural science has, Quine holds, its own preferred repertoire of types of objects to the existence of which it is committed. Ontology is then not the meta-level study of the ontological commitments or presuppositions embodied in the different natural-scientific theories. Ontology, for Quine, is rather these commitments themselves.

Quine fixes upon the language of first-order logic as the medium of canonical representation in whose terms these commitments will become manifest. He made this choice not out of dogmatic devotion to some particular favoured syntax, but rather because he holds that the language of first-order logic is the only really clear form of language we have. His so-called 'criterion of ontological commitment' is captured in the slogan: *to be is to be the value of a bound variable*. This should not be understood as signifying some reductivistic conception of being – as if to exist would be a merely logico-linguistic matter – something like a mere *façon de parler*. Rather, it is to be interpreted in practical terms: to determine what the ontological commitments of a scientific theory are, it is necessary to determine the values of the quantified variables used in its canonical (first-order logical) formalization.

One problem with this approach is that the objects of scientific theories are discipline-specific. How, then, are we to approach the issue of the compatibility of these different sets of ontological commitments. Various different solutions have been suggested for this problem, including reductionistic solutions, based on the conception of a future perfected state of science captured by a single logical theory and thus marked by a single, consistent and exhaustive set of ontological commitments.

At the opposite extreme is a relativistic approach, which renounces the very project of a single unitary scientific world view (and which might in principle invite into the mix the ontological commitments of non-scientific world views, as embraced for example by different religious cultures). The 'external' question of the relations between objects belonging to different disciplinary domains falls out of bounds for an approach along these lines.

The adequatist approach to ontology stands in contrast to both of these perspectives, holding that the issue of how different scientific theories (or how the objects described by such theories) relate to each other is of vital importance, and can be resolved in a way which does justice to the sciences themselves. For Quine, the best we can achieve in ontology lies in the quantified statements *of particular theories*, theories supported by the best evidence we can muster. We have no extra-scientific way to rise above the particular theories we have and to harmonize and unify their respective claims. This implies also that philosophers lack authority to interfere with the claims and methods and empirical data of scientists. In the current age of information-driven science, however, tasks of the sort which were in earlier epochs addressed by philosophical ontologists, and which in the era of Quine and Carnap (and of their precursors in the Vienna Circle) were seen as falling in the province of logicians, are now being addressed by computer scientists.

On the Way to Computational Ontology

As scientists must increasingly rely on the use of computer systems to absorb the vast quantities of information with which they are confronted, and as computers are being applied to the storage and integration of multiple different kinds of scientific data, computer scientists are being called upon to address problems which in earlier times had been addressed by those with philosophical training.

In a development hardly noticed by philosophers, the term 'ontology' has hereby gained currency in the field of computer and information science, initially through the avenue of Quine, whose work on ontological commitment attracted the attention of researchers in artificial intelligence, such as John McCarthy[8] and Patrick Hayes,[9] and from the programming world, such as Peter Naur. As McCarthy expressed it in 1980, citing Quine in his use of 'ontology', builders of logic-based intelligent systems must first 'list everything that exists, building an *ontology* of our world'. In 1999, a new wave of computationally oriented ontology developments began in the world of bioinformatics with the creation of the Gene Ontology (GO).[10]

The GO addresses the task of solving the problem of data integration for biologists working on so-called 'model organisms' – genetically tailored mice or fish or other organisms – which are used in experiments designed to yield results which will bring consequences for our understanding of human health and of the effects of different kinds of treatment. The problem faced by the GO's authors turned on the fact that each group of model organism researchers had developed its own idiosyncratic vocabularies for describing the phenomena revealed in their respective bodies of data. Moreover, these vocabularies were in turn not consistent with the vocabularies used to describe the human health phenomena to the understanding of which their research was directed. Different groups of researchers used identical labels but with different meanings, or they expressed the same meaning using different names. With the explosive growth of bioinformatics, ever more diverse groups became involved in sharing and translating ever more diverse varieties of information at all levels, from molecular pathways to populations of organisms, and the problems standing in the way of putting this information together within a single system began to increase exponentially. By providing a solution to these problems in the form of a common, species-neutral, controlled vocabulary covering the entire spectrum of biological processes, the GO has proved tremendously successful (see Figure 7.2), and is almost certainly the first real demonstration case of the advantages brought by ontological technology in supporting the integration of data for scientific purposes.

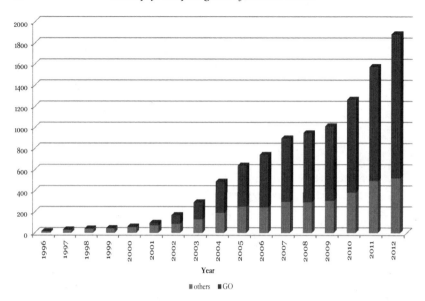

Figure 7.2: Number of articles on ontology or ontologies in PubMed/MEDLINE.
Updated from: O. Bodenreider and R. Stevens, 'Bio-Ontologies: Current Trends and
Future Directions', *Briefings in Bioinformatics*, 7:3 (2006), pp. 256–74, on p. 265.

As the GO community has discovered, however, the success of an ambitious
ontology initiative along these lines faces a constant need to identify and resolve
the inconsistencies which arise as its terminological resources are expanded
through the contributions of multiple groups engaged in different kinds of biological research.

Initially, such incompatibilities were resolved by the GO – and by the authors
of the new ontologies which had grown up in its wake – on a case-by-case basis.
Gradually, however, it came to be recognized in the field of bio-ontologies that
the provision of common reference ontologies – effectively, shared taxonomies of
entities – might provide significant advantages over such case-by-case resolution.
An ontology is in this context a dictionary of terms formulated in a canonical
syntax and with commonly accepted definitions designed to yield a lexical or
taxonomical framework for knowledge representation which can be shared by
different information systems communities. More ambitiously, an ontology is
a formal theory within which not only definitions but also a supporting framework of axioms are included (perhaps the axioms themselves provide implicit
definitions of the terms involved). The methods used in the construction of
ontologies thus conceived are derived, on the one hand, from earlier initiatives
in database management systems. But they also include methods similar to those
employed in philosophy (as described already in Hayes[11]), including the methods used by logicians when developing formal semantic theories.

Upper-Level Ontologies

The potential advantages of ontology for the purposes of information management are obvious. Each group of data analysts would need to perform the task of making its terms and relations compatible with those of other such groups only once – by calibrating its results in terms of a single canonical backbone language. If all databases were calibrated in terms of just one common ontology (a single consistent, stable and highly expressive set of category labels), then the prospect would arise of leveraging the thousands of person-years of effort that have been invested in creating separate database resources in fields such as biochemistry or computational biology such a way as to create, in more or less automatic fashion, a single integrated knowledge base.

The obstacles standing in the way of the construction of a single shared ontology in the sense described are unfortunately prodigious, ranging from technical difficulties in choice of appropriate logical framework,[12] difficulties in coordination of different ontology authoring communities, difficulties which flow from the entrenched tendencies of many computer scientist communities to react negatively to the idea of reusing already created computational artefacts and to prefer much rather to create something new for each successive customer.[13] Added to this are the difficulties which arise at the level of adoption. To be widely accepted an ontology must be neutral as between different data communities, and there is, as experience has shown, a formidable trade-off between this constraint of neutrality and the requirement that an ontology be maximally wide-ranging and expressively powerful – that it should contain canonical definitions for the largest possible number of terms. One solution to this trade-off problem is the idea of a top-level ontology, which would confine itself to the specification of such highly general (domain-independent) categories as: time, space, inherence, instantiation, identity, measure, quantity, functional dependence, process, event, attribute, boundary, and so on. The top-level ontology would then be designed to serve as common neutral backbone, which would be supplemented by the work of ontologists working in more specialized domains on, for example, ontologies of geography, or medicine, or ecology, or law. An ambitious strategy along these lines is currently being realized in the domains of biology and biomedicine,[14] and it is marked especially by the adoption of a common top-level ontology of relations, which provides the common formal glue to link together ontologies created by different communities of researchers.[15]

Some Critical Remarks on Conceptualizations

Drawing on the technical definition of 'conceptualization' introduced by Genesereth and Nilsson in their *Logical Foundation of Artificial Intelligence*,[16] in 1993 Tom Gruber introduced an influential definition of 'ontology' as 'the specification of a conceptualization'.[17] One result of Gruber's work was that it became common in computer circles to conceive of 'ontology' as meaning just 'conceptual model'. *Applied Ontology*, the principal journal of the ontology engineering field, accordingly has the subtitle *An Interdisciplinary Journal of Ontological Analysis and Conceptual Modeling*.

For Gruber, 'A conceptualization is an abstract, simplified view of the world that we wish to represent for some purpose. Every knowledge base, knowledge-based system, or knowledge-level agent is committed to some conceptualization, explicitly or implicitly.'[18] The idea, here, is as follows. As we engage with the world from day to day we use information systems, databases, specialized languages and scientific instruments. We also buy insurance, negotiate traffic, invest in bond derivatives, make supplications to the gods of our ancestors. Each of these ways of behaving involves, we can say, a certain conceptualization. What this means is that it involves a system of concepts in terms of which the corresponding universe of discourse is divided up into objects, processes and relations in different sorts of ways. Thus in a religious ritual setting we might use concepts such as *salvation* and *purification*; in a scientific setting we might use concepts such as *virus* and *nitrous oxide*; in a storytelling setting we might use concepts such as *leprechaun* and *dragon*. Such conceptualizations are often tacit; that is, they are often not thematized in any systematic way. But tools can be developed to specify and to clarify the concepts involved and to establish their logical structure, and thus to render explicit the underlying taxonomy. An 'ontology' in Gruber's sense is, then, the result of such a clarification employing appropriate logical tools.

Ontology, for Gruber and for the many computer scientists who have followed in his wake, thus concerns itself not with the question of ontological realism, that is with the question whether its conceptualizations are *true of* some independently existing reality. Rather, it is a strictly pragmatic enterprise. It starts with conceptualizations, and goes from there to the description of corresponding domains of objects – often themselves confusingly referred to as 'concepts' – which are not real-world entities but rather abstract nodes in simplified computer models created for specific application purposes.

Against this background, the project of developing a top-level ontology, a common ontological backbone, begins to seem rather like the attempt to find some highest common denominator that would be shared in common by a plurality of true and false theories. Seen in this light, the principal reason for the failure of so many attempts to construct shared top-level ontologies lies precisely in the fact that these attempts were made on the basis of a methodology which treated all application domains on an equal footing. It thereby overlooked the degree to which the different conceptualizations which serve as inputs to ontology are likely to be not only of wildly differing quality but also mutually inconsistent.

The Open Biomedical Ontologies (OBO) Foundry,[19] which is one promising attempt to create an interoperable suite of ontologies sharing a common top-level ontology, seems to be succeeding in this respect primarily because it is restricted to domains where an independently existing reality – of biological and biomedical entities studied by science – serves as a constraint on the content of the ontologies included within the OBO framework. Ontology for the OBO Foundry, in other words, is not a matter of conceptual modelling.[20]

8 ONTOLOGY, ITS ORIGINS AND ITS MEANING IN INFORMATION SCIENCE

Jens Kohne

Introduction

Ontology – in Aristotelian terms the science of being *qua* being – as a classical branch of philosophy describes the foundations of being in general. In this context, ontology is general metaphysics: the science of everything.[1] Pursuing ontology means establishing some systematic order among the being, i.e. dividing things into categories or conceptual frameworks. Explaining the reasons why there are things or even anything, however, is part of what is called special metaphysics (theology, cosmology and psychology). If putting things into categories is the key issue of ontology, then general structures are its main level of analysis. To categorize things is to put them into a structural order. Such categorization of things enables one to understand what reality is about. If this is true, and characterizing the general structures of being is a reasonable access for us to reality, then two kinds of analysis of those structures are available: (i) realism and (ii) nominalism.

In a realist (Aristotelian) ontology the general structures of being are understood as a kind of mirror reflecting things in their natural order. Those categories, as they are called in realism, then represent or show the structure of being. Ontological realism understands the relation between categories and being as a kind of correspondence or mapping which gives access to reality itself.

The converse is true of ontological nominalism. Categories, or rather conceptual schemes, as they are called in nominalism, do not represent the structure of being at all: behind those structures there is no further reality. The only reality to which we have access is the general conceptual schemes by which we give reality a structure. This means that conceptual schemes frame the structure of being as reality. So from the nominalist position the direction of the representation between general structure and being is the converse of the direction of representation in realism.

Of course this fundamental difference between the two ontological theories is the result of different ontological concepts of being. In the nominalist sense, being is mind-dependent, whereas the realist claims that being exists independently of the mind. That entails an epistemological commitment to a realist approach which does not take place in the nominalist position, a commitment to an accessible knowledge of entities independent of the mind that represents them.

Now, if information science too uses ontology in the sense of examining general structures[2] then the question arises as to what kind of status those structural frameworks have: do they represent a mind-independent reality or are they bare structures of mind-dependent representations?

This essay elaborates on the problem and outlines the consequences these two ontological positions have for the usage of ontology in information science.

The Origins of Ontology

If we take a look at the history of philosophy, there are three names for the discipline called ontology or three names, rather, that relate to ontology: metaphysics, first philosophy and, after Glocenius's *Lexicon Philosophicum* and Lorhardus's *Theatrum philosophicum* in 1613, ontology.[3] The origins of metaphysics as a discipline can be found in an Aristotelian treatise of the same name. Although Aristotle himself never called the treatise by that title – he used to call the discipline First Philosophy – in it he describes in detail what metaphysics is all about. Aristotle gives two accounts of metaphysics: the departmental discipline which identifies first causes, in particular God, and, second, the universal science whose task it is to examine being *qua* being. That implies an examination of the general features of everything that there is: existence in general is the subject matter here. Pretty clearly, Aristotle needs a very general instrument if he wants to study the pure existence of a thing. Those instruments he calls *categories*. The Aristotelian categories classify the nature of being of everything there is, so that we are able to categorize all things into general kinds. In doing so we are able to identify all the kinds of beings that there are, which enables the preparation of a general map showing the most general structure of everything there is. In this sense metaphysics is the science that characterizes the most general structures of what there is.[4]

As Aristotle sees things, this is the enterprise of metaphysics: the identification of first causes and mapping the most general structures of reality. But the history of philosophy – especially in the seventeenth and eighteenth centuries – displays the difficulty of thinking of both perspectives of metaphysics, the departmental discipline of first causes and the universal science of the most general structures of being, as one consistent discipline. Since the two different perspectives of the metaphysical enterprise indeed have something in common – the general search for the nature of being – philosophers decided to redefine the enterprise of metaphysics as *general metaphysics* and *special metaphysics*. 'General metaphysics', which has since Glocenius's and Lorhardus's time been given the new name of ontology, stands for the universal discipline of categorizing reality, and the study of first causes which entails the departmental disciplines of cosmology, rational psychology and natural theology is now signified by the term 'special metaphysics'.[5]

After the redefinition and expansion of Aristotelian metaphysics, Kantian philosophy gave new incentives to metaphysics. If we leave special metaphysics aside, the consequences of those idealistic incentives were a dramatic change in the ontological realm. As we have seen, since Aristotle the metaphysician has been in search of the nature of being. That means she wants to know the foundations of being, and her instruments to look for those foundations are the descriptions of the most general structural coherences between the things that there are by means of categories. To have those categories is to have the most general structure of reality. But if this is the ontological enterprise, the assumption needs to be true that there is a cognitive access to the nature of being so that metaphysicians can describe it by means of categories. So the (old) crucial question is: does a reality exist independent of my mind and my representations of it? Second, if so, do I have cognitive access to the independent reality? If the answer to these questions is in the negative, an Aristotelian account of metaphysics is no longer possible.[6]

The Kantian, idealistic, answer to the last question is definitely no. There is no cognitive access available to the independent reality outside my mind. The conditions of my mind constitute the reality I am living in because

> to think of anything external to my cognitive faculties, I must apply concepts that represent the thing as being some way or other, as belonging to some kind or as characterized in some way; but, then, what I grasp is not the object as it really is independently of my thought about it. What I grasp is the object as I conceptualize or represent it, so that the object of my thought is something that is, in part at least, the product of the conceptual or representational apparatus I bring to bear in doing the thinking. What I have is not the thing as it is in itself, but the thing as it figures in the story I tell of it or the picture I construct of it.[7]

Instead of quitting the ontological enterprise now, Kant changed its purpose. In place of describing the nature of being, a characterization of the conceptual framework which constitutes reality is needed now. That means ontology is no longer interested in a direct identification and characterization of reality via categories but in a conceptualization of how reality is represented in my mind or rather through my mind, inasmuch as the difference between the Aristotelian ontology and the idealistic one is between two levels of considering reality. As Loux comments, 'An inquiry into the structure of human thought is, however, something quite different from an inquiry into the structure of the world thought is about'.[8]

Aristotelian ontology is a first-order ontology because it is concerned with what reality is really about, and because of that we can call it realism.[9] Unlike realism, idealistic ontology is not concerned with what reality is really about – the nature of being – because it is inaccessible to us. So the idealistic ontologist is slipping in a second-order or meta-level in order to do an ontological enterprise. On this second-order level the idealistic ontologist moves away from questions about reality itself and turns to questions about the representation of reality. Since representation happens via concepts this ontological enterprise is now only concerned with the description of the most general conceptual framework by which the representation of reality is structured. This second-order ontology we can name nominalism, because its aim is only a characterization of the most general conceptual structure representing things without touching on the crucial point of whether those structures represent anything outside or independent of the concepts. The nominalist has restricted her ontological efforts to an analysis of the way we use concepts in representing things, i.e. talking, thinking and so forth.[10] Alternatively, as Loux (quoted above) says, doing ontology in this sense is 'an inquiry into the structure of human thought'.

But why should there be such a fundamental difference between a realist and a nominalist way of doing ontology? The difference seems to be rather gradual. 'Of course, if one believes that the structure of our thought reflects or mirrors the structure of the world, then one might claim that the results of the two inquiries must be the same.'[11]

It seems as if both protagonists, nominalist and realist, in the end do the same thing: the categories mirror the fundamental structures of reality and concepts structure the representations with which we reflect reality.

But the point is this: categories are not concepts. Aristotelian categories are the fundamental structures of reality, a kind of ontological glue as David Armstrong would say.[12] They are not representing anything at all in a non-trivial sense. In fact, they are themselves a fundamental part of reality, insofar as the two ontological positions are not doing the same job in different ways. Indeed, there is a fundamental gap between these two accounts of reality which cannot be bridged, because the central claim of the nominalist position is 'that our thought about the world is always mediated by the conceptual structures in terms of which we represent that world'.[13]

Some Consequences for the Usage of Ontology in Information Science

Now, what kind of consequences follow from that short consideration about the foundations of ontology? If we first of all take a quick look at the definition of ontology in information science, we notice that there seems to be an analogy between that and the philosophical definition elaborated above.[14] In information science, too, ontology describes the examination of general structural frameworks or conceptualizations and, as in philosophy, those frameworks are of representations. But what do those frameworks represent? The answer we find in Gruber, Smith or Hesse is: various kinds of knowledge.[15]

Categorizations in computer science, then, seem to be necessary to deal with specified contents of knowledge so they can be suitably employed in user-defined applications. In this context ontology means establishing special derivation rules for managing knowledge machinably.[16] If, however, the knowledge that computer science wants to manage machinably via special derivation rules is knowledge about reality, then we are playing the realism/nominalism game, with all the consequences stated above.

Here again the question is, what is the status of those representations? Are they mind-dependent, representing only the conceptual frameworks a machine works with in order to manage a special content of knowledge?[17] Or are they mind-independent categories representing general structures of being which computer science is now trying to translate into quasi machine-readable derivation rules? If the latter is true, machines – principally – could generate knowledge which is mind-independent, i.e. giving access to the nature of reality.

But if doing ontology in computer science means the former, then the knowledge which is generated by machines could only reflect a reality within the constraints of their system conditions. That means we are unable to gain any new insights to add to the things already known. It's like being on a treadmill: you are spending energy, but you are getting nowhere or, to put the same point in a slightly different way, what is happening is only a transfiguration of the virtual. Outside the system conditions there is no further reality available any longer: you are entering the virtual reality.

9 SMART QUESTIONS: STEPS TOWARDS AN ONTOLOGY OF QUESTIONS AND ANSWERS

Ludwig Jaskolla and Matthias Rugel

Introduction

The present essay is based on research funded by the German Ministry of Economics and Technology and carried out by the Munich School of Philosophy (Prof. Godehard Brüntrup) in cooperation with the IT company Comelio GmbH. It is concerned with setting up the philosophical framework for a systematic, hierarchical and categorical account of questions and answers in order to use this framework as an ontology for software engineers who create a tool for intelligent questionnaire design.

In recent years, there has been considerable interest in programming software that enables users to create and carry out their own surveys. Considering the, to say the least, vast amount of areas of applications these software tools try to cover, it is surprising that most of the existing tools lack a systematic approach to what questions and answers really are and in what kind of systematic hierarchical relations different types of questions stand to each other. The theoretical background to this essay is inspired Barry Smith's theory of regional ontologies.[1]

The notion of ontology used in this essay can be defined by the following characteristics: (1) The basic notions of the ontology should be defined in a manner that excludes equivocations of any kind. They should also be presented in a way that allows for an easy translation into a semi-formal language, in order to secure easy applicability for software engineers. (2) The hierarchical structure of the ontology should be that of an *arbor porphyriana*. That is to say, it should be presented in a hierarchical tree of categories branching disjunctively. For every two levels of the hierarchy that are directly connected one should be able to name a *differentia specifica* which clearly distinguishes the lower level of the tree from the higher level.[2]

Keeping these requirements in mind, we will discuss the basic notions of our ontology in the following section of this essay (pp. 92–4). The third section (pp. 94–5) will be concerned with the ontological status of the structures as developed in part two. The fourth section (pp. 95–7) will give an outline of the complete *arbor porphyriana* for our ontology, whereas the fifth section (p. 97) will try to give a glimpse of what has to be done in order to set up a full-blown ontology of questions and answers.

Basic Definitions for an Ontology of Questions

Our central intuition establishing a systematic account for questions and answers was that an ontology for the purposes of software engineering cannot be deduced from a mere collection of different types of questions we are using in ordinary language. In Figure 9.1 we give a short visual example of what an ontology deduced from ordinary language could look like:

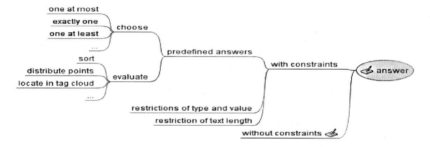

Figure 9.1: Representation of an ontology deduced from ordinary language.

The main reason for this intuition is the following consideration: questions that are used in surveys are structurally different from questions we pose in ordinary language. The difference originates from the fact that in ordinary language we are talking to particular individuals answering our questions. When considering questions in surveys we are confronted with a huge amount of answers that can only be handled by means of statistical analysis. So whereas we are just asking things in ordinary language, when posing questions in a survey we must take into account that these questions have to provide data that is evaluable by the means of statistical analysis. So we tried to construe our ontology of questions in accordance to the statistics used by the social sciences. But our ontology does not want to be completely revisionary. The types of questions used in ordinary language should rather be used to ensure that the ontology covers the whole area of ordinary asking and answering. In analogy to Popper's thesis that a scientific theory should be accounted for as false if there are empirical facts contradicting the theory, we want to consider an ontology of questions as false, if there are types of questions used in ordinary language that are not covered by this specific ontology. Moreover, taking into account Popper's even stronger thesis that scientific theories can never be positively verified, we are stating that there is in principle no positive verification for an ontology of questions. Thus, to use ordinary types of questions as a test for possible falsification seems to be an important aspect concerning the creation of an ontology of questions.

What can be deduced from this consideration that the question in a survey is essentially tied to the evaluation? Surely, it will depend on the structure of the answer this particular question is related to. Thus, in a first, approximate 'definition' we can state the following:

(Proposal) A question is dependent on the structure of the answer; in fact being nothing more than the syntactically modified (in a grammatical correct way) answer.

We need to spell out that first step towards a definition in terms of statistics: in the social sciences the set of all subjects of investigation in one survey is called the *population* of that survey. So in classical surveys the population covers all the things that are surveyed and all the people partaking in a survey. This seems to be rather undesirable because there are two different kinds of entities that are covered by this definition – the objects of the survey, and the subjects partaking in the survey. Due to this fact, we decided to distinguish these two kinds of entities in our definition of population.[3] Here it is:

(Definition 1) A subject-population covers the entities that can be the object of one specific survey AND are able to partake in this survey themselves. On the other hand, an object-population covers all the entities that can only be the object of this survey.

To spell this definition out in an example: we want to ask German students about their living conditions in dorms. The subject-population of this survey would be the set of people enrolled at a German university living in dorms. The object-population would consist of the dorms, especially the properties of the dorms that are tied to the living conditions in these dorms.

It is possible that subject- and object-population are identical, when for example a group of people is asked about its religious beliefs. But that seems only to be the special case of Definition 1 presented above. Considering our ontology of questions, we can state that a question in a survey is tied to one and only one population.

A question attempts to find out something about one specific *attribute* of a member of the object-population. In our preceding example that attribute could be the heating of the dorm building. The attribute has itself different *characteristics*. These characteristics define the structure, how the attribute is presented in the survey. Thus, we can easily see that if we want to construe our ontology in accordance with the means of statistical analysis, the different aspects of the evaluation of the question will have to occur in the structure of the characteristics of a particular attribute. In the preceding example, characteristics of the attribute 'heating' could be the different means of heating used in the dorm building – 'coal', 'gas', 'renewable energy', and so on.[4]

These are the essential aspects of questions in the context of the social sciences. In the rest of this section, we want to put these different aspects together to define the notion of a question in our ontology.

(Definition 2) A question is the syntactical modified structure of characteristics of a particular attribute concerning a subject-population.

We want to spell out this definition in another example. Consider a question querying the marital status of the people of Germany, and consider the characteristics of the marital status to be 'married', 'divorced', 'widowed' and 'single'. Due to Definition 2 we are able to describe the question as a vector bearing the structure presented here:

$$\textbf{Question}_{\textit{Marital Status},\,\Omega}(\omega)=[\mathit{Q}]*\begin{bmatrix} \textit{married} \\ \textit{divorced} \\ \textit{widowed} \\ \textit{single} \end{bmatrix}(\omega)$$

In the following section of this essay, we will be concerned with the ontological status of object- and subject-populations, before we try to set up an *arbor porphyriana* for different kinds of questions.

Realism versus Anti-Realism for Populations

The philosophical question concerning the ontological status of subject- and object-populations is an important one for the ontology presented in this essay. Two approaches seem promising:

One could think of subject- and object-populations as containing no entities that are different in principle. Due to this approach all aspects of an entity that can be queried in a survey are only 'mental attitudes' the people partaking in the survey hold to be true about the world. And thus, we have no direct knowledge about the real structure of the world surrounding us. In the following considerations we want to call that position 'anti-realism for populations'.

By contrast, 'realism for populations' states that there are two distinct areas of the world that can be queried in a survey. One area concerns the 'concepts' people have about certain features of the objects surrounding them, for example when we want to know how trustworthy someone rates a particular politician. And the other area queries certain objective features of the world, for example when we want to find out how much, on average, a particular airline is behind schedule at a certain airport. Because this approach allows for objective features of the world to be known by people, it will be called 'realism for populations'.

The main advantage of the anti-realist approach seems to be that anti-realism would make our ontology much simpler. And because of the fact that simplicity should be sought in an ontology for any one part of reality, we need to consider whether anti-realism for populations is a genuine alternative to the realist approach that corresponds to the common-sense understanding of populations in the preceding definitions.

In our opinion, there are two main reasons to reject anti-realism in an ontology for questions. First, it seems that the anti-realist approach only pushes back the problematic distinction of 'concepts' and objective knowledge into the mind of the person partaking in the survey. This is because anti-realists have to make a distinction between states of the mind that are accessible to others (see objective knowledge) and that are not (see mere concepts). Otherwise anti-realism would make our ordinary-life assumption of these two areas obsolete. Second, anti-realism seems to undermine our intuition that we are asking for objective features of the world when asking certain questions. We tried to show that in the example of flight delays. What we want to know are objective features (in this case: the exact time of arrival stored in a computer), and not what the flight attendant remembers.

To summarize, we think that, in contrast to first impressions, realism for populations is simpler than anti-realism when trying to set up an ontology for questions. In principle, however, our ontology is neutral in the realism-antirealism debate.[5] In the following section of this essay, we will try to describe this ontology, but need to take in account that the realistic approach seems to fit better with our ordinary understanding of the world.

An *Arbor Porphyriana* for an Ontology of Questions

There are two main aspects that need to be discussed when setting up the ontology of questions: first, we think that it is important to distinguish (mathematically) continuous and non-continuous structures of characteristics for an attribute.

The main intuition of the second aspect is to use the metrical features of statistical analysis in describing the structure of characteristics for a particular attribute. Again, we try to stay as close to the exact definitions of statistics as possible. We thus distinguish structures that are scaled (i) nominally from those that are scaled (ii) ordinally and those that are scaled (iii) metrically.[6]

Text concerning (i): Nominally scaled structures are those where only the identity or non-identity of characteristics matters. According to our research, querying qualitative characteristics (for example the preferred colour of a car) amounts to querying an attribute whose characteristics are nominally scaled.

Text concerning (ii): In contrast to nominal scaling, ordinal scaling allows for different answers to a question to be compared by the mathematical relations of 'greater as (>)' and 'smaller as (<)'. One would want to use this scaling when, for example, querying the number codes on identity cards.

Text concerning (iii): Metric scaling adds the feature of being able to measure the difference between the characteristics of an attribute exactly, for example when querying the different temperatures at a certain place throughout the year. This scaling allows the calculation of the arithmetic mean as a special feature.[7]

When putting all these considerations together, we get the following *arbor porphyriana* for an ontology of questions (see Figure 9.2). The *differentiae specificae* are set in italics and marked by the prefix 'diff'. The text in bubbles, on the other hand, marks the different categorical distinctions that appear in this ontology of questions.

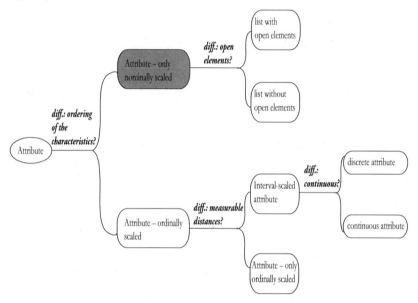

Figure 9.2: Representation of an *arbor porphyriana* for an ontology of questions.

(1) An example for a list without open elements could be a definite set of colours, when asking for the preferred colour of a car. (2) An example for a list with open elements could be the same as (1) but the set would also contain one open element, where one could put in a preferred colour if it is not on the list already. (3) An example for an ordinally scaled attribute has been discussed above (the number codes of IDs). (4) Discrete attributes could be something like asking how many children live in a household. (5) And to conclude the tree, an example for a continuous attribute could be asking for someone's age group ('Are you [10–20] or [20–30], and so on).

This new ontology includes some remarkable advantages when compared to the hitherto used ordinary-language ontology that was briefly outlined in the second section of this essay (p. 92). At first, we can record that this ontology fulfils the requirements that were formulated in the introductory part of this essay: there are no equivocations and it is a complete *arbor porphyriana*. Second, this ontology uses the language of statistical analysis and is therefore fit for being used as a systematization for software tools that create questionnaires. Third, concerning the question of computability, our ontology is able to distinguish between the categorical status of a question and the representation of the same question in the user interface. This should enhance the computational applicability of the ontology that is stated in this essay.

Prospects on What Has to Be Done in Addition

Clearly there is still some work to be done before one can give a definite classification for types of questions. In the concluding part of this chapter, we want to give a short outline of the method we are presently using to give a complete description of ordinary question types.

A definite description of a question starts by ranking a question in one end of our tree. Then it seems necessary to further determine the functionalities for this particular question type. One of these functionalities could be that the characteristics are presented in a randomized order.

After that we are to ascribe one or more selection rules to the question. These rules describe in which ways a question can be answered, for example by being able to evaluate the characteristics.

At the end of this process, we will be concerned with the graphical presentation of the question. But this seems only a somewhat arbitrary step in the definite description of a question type. These further steps should be enough to give a description of a question in a manner that fulfils the requirements of an ontology and is exact enough to be easily applicable in software-engineering processes.

10 SOPHISTICATED KNOWLEDGE REPRESENTATION AND REASONING REQUIRES PHILOSOPHY

Selmer Bringsjord, Micah H. Clark and Joshua Taylor

What is Knowledge Representation and Reasoning?

What is knowledge representation and reasoning (KR&R)? Alas, a thorough account would require a book,[1] or at least a dedicated, full-length paper,[2] but here we shall have to make do with something simpler. Since most readers are likely to have an intuitive grasp of the essence of KR&R, our simple account should suffice. The interesting thing is that this simple account itself makes reference to some of the foundational distinctions in the field of philosophy. These distinctions also play a central role in artificial intelligence (AI) and computer science.

To begin with, the first distinction in KR&R is that we identify knowledge with knowledge that such-and-such holds (possibly to a degree), rather than knowing how. If you ask an expert tennis player how he manages to serve a ball at 130 miles per hour on his first serve, and then serve a safer, topspin serve on his second should the first be out, you may well receive a confession that, if truth be told, this athlete can't really tell you. He just does it; he does something he has been doing since his youth. Yet, there is no denying that he knows how to serve. In contrast, the knowledge in KR&R must be expressible in declarative statements. For example, our tennis player knows that if his first serve lands outside the service box, it's not in play. He thus knows a proposition, conditional in form. It is this brand of declarative statement that KR&R is concerned with.

At some point earlier, our tennis player did not know the rules of tennis. Suppose that for his first lesson, this person walked out onto a tennis court for the first time in his life, but that he had previously glimpsed some tennis being played on television. We can thus imagine that before the first lesson began, our student believed that a serving player in tennis is allowed three chances to serve the ball legally. This belief would have of course been incorrect, as only two chances are permitted. Nonetheless, this would be a case of a second attitude directed toward a proposition. The first attitude was knowledge, the second mere belief.

Knowledge-based systems (KBSs), then, can be viewed as computational systems whose actions through time are a function of what they know and believe. Knowing that his first serve has landed outside the service box on the other side of the net from him, our (educated) tennis player performs the action of serving for a second time, and as such performs as a KBS. A fully general and formal account of KBSs can be found elsewhere.[3] There are numerous algorithms designed to compute the functions in question, but in the present essay we shall be able to rest content with reference to but a few of them.

The Nature of Philosophy-less KR&R

In this section, after introducing the basic machinery of elementary extensional and intensional logic for purposes of KR&R carried out in the service of building KBSs, we present our characterization of the dividing line between philosophy-less KR&R versus philosophy-infused KR&R.

Overview of Elementary Extensional and Intensional Logic for KR&R

Propositions can be represented as formulas in formal languages. For example, in the present case, we might use a simple formula from the formal language $\mathcal{L}_{PC}$ of the *propositional calculus*, which allows propositions to be expressed as either specific propositional variables such as $p_1, p_2, \ldots$ (or mnemonic replacements thereof, e.g. h for p_2 when we want to represent the proposition that John is happy), or as formulas built from the p_i and the familiar Boolean connectives: $\neg\varphi$ ('not φ'), $\varphi\vee\psi$ ('φ or ψ'), $\varphi\wedge\psi$ ('φ and ψ'), $\varphi\to\psi$ ('if φ then ψ'), $\varphi\leftrightarrow\psi$ ('φ if and only if ψ').[4] For example, letting *out* represent the proposition that the ball lands outside the service box, and *play* that the ball is in play, *out* $\to \neg$ *play* represents the above conditional. If a knowledge-based system knows *out* and this conditional, it would of course be able to infer $\neg$ *play*, that the ball is not in play. Its reasoning would be deductive in nature, using the well-known rule of inference *modus ponens*. To use the standard provability relation $\vdash_X$ in knowledge-based/logic-based AI and cognitive science, where the subscript X is a placeholder for some particular proof calculus, we would write

$$\{out,\ out \to \neg\ play\} \vdash_X \neg play$$

to express the fact that the ball's being out of play can be proved from the formulas to the left of $\vdash$. So here we have a (painfully!) simple case of a KBS, powered by KR&R in action.

Some discussion concerning candidate proof calculi for X is necessary. Due to lack of space, we must leave aside specification of each of the myriad possibilities, in favour of a streamlined approach. This approach is based upon the incontestable fact that there clearly is a *generic* conception of fairly careful deductive reasoning according to which some lines of linear, step-by-step inference can be accepted as establishing their conclusions with the force of proof, even though detailed definition of particular calculus X, and use thereof, is absent. This streamlined approach works because the step-by-step sequence is such that each step from some set $\{\varphi_1,..., \varphi_n\}$ to some inferred-to formula Ψ can be quickly seen, with a small amount of mental energy, to be such that *it is impossible that each φ_i hold, while Ψ does not.*[5] What follows is an example of such a sequence, couched in natural language; the sequence establishes with the force of proof that 'from 'Everyone likes anyone who likes someone' and 'Albert likes Brian' it can be inferred that 'Everyone likes Brian'. (Most people see that it can be inferred from this pair of statements that 'Everyone likes Albert', but are initially surprised that 'Everyone likes Brian' can be derived in a bit of a recursive twist.)

1	Everyone likes anyone who loves someone.	assumption
2	Albert likes Brian.	assumption
3	Albert likes someone.	from 2
4	Everyone likes Albert.	from 1, 3
5	Brian likes Albert.	from 4
6	Brian likes someone.	from 5
7	Everyone likes Brian.	from 1, 6

Despite opting for what we have called a streamlined approach to provability in the present essay, we would be remiss if we failed to point out that the format that best coincides with how professionals in those fields based on deductive reasoning actually construct proofs and disproofs is clearly something quite like 'Fitch-style' *natural deduction*, with which many readers will be acquainted.[6] In this kind of 'proof' calculus, which aligns with the deductions written in relevant professional books and papers (in computer science, mathematics, logic, etc.), each of the truth-functional connectives, and the quantifiers (see below), has a pair of corresponding inference rules, one for introducing the connective, and one for eliminating the connective. One concrete possibility for a natural-deduction calculus is the 'human-friendly' one known as $\mathcal{F}$.[7] Another possibility is the natural-deduction-style proof calculus used in the Athena system.[8] We make use of the Athena system below, but don't use or specify its proof calculus.

The propositional calculus is rather inexpressive. Most of what we know cannot be represented in $\mathcal{L}_{PC}$ without an unacceptably large loss of information. For example, from the statement 'Albert likes Brian', we can infer that 'Albert likes someone'. We might attempt to represent these two statements, respectively, in $\mathscr{L}_{PC}$ as, say, A and $A_{someone}$. Unfortunately, this representation is defective, for the simple reason that by no acceptable rule of deductive inference can $A_{someone}$ be deduced from A. The problem is that $\mathscr{L}_{PC}$ cannot express quantification in formulas such as 'Albert likes someone', and so lacks inference rules such as *existential introduction* (which formally obtains the intuitive result above).[9] The machinery of quantification (in one simple form), and this particular rule of inference, are part of *first-order logic* , whose formal language is $\mathscr{L}_{FOL}$. The alphabet for this language reflects an increase over that for the propositional calculus, to include:

identity	=	the identity or equality symbol;
connectives	¬, ∨, ...	now familiar to you, same as in $\mathscr{L}_{PC}$;
variables	$x, y, ...$	variables ranging over objects;
constants	$c_1, c_2, ...$	you can think of these as proper names for objects;
relations	$R, G, ...$	used to denote properties, e.g., W for *being a widow*;
functions	$f_1, f_2, ...$	used to refer to functions;
quantifiers	∃, ∀	∃ says 'for some ...'; ∀ says 'for every ...'

Predictable *formation rules* are introduced to allow one to represent propositions like the 'Everyone likes anyone who likes someone' one above. In the interests of space, the grammar in the question is omitted, and we simply show 'in action' the kind of formulas that can be produced by this grammar, by referring back to the Albert–Brian example. We do so by presenting here the English-based sequence from above in which natural language is replaced by suitable formulas from $\mathscr{L}_{FOL}$. Recall that this sequence, in keeping with the streamlined approach to presenting provability herein, is something that qualifies as an outright proof. In addition, the reader should rest assured that automated theorem proving technology of today can instantly find a proof of line 7 from lines 1 and 2.[10]

1	$\forall x \forall y [(\exists z\ Likes\ (x,z)) \rightarrow Likes\ (y,x)]$	assumption
2	$Likes(a, b)$	assumption
3	$\exists x\ Likes(a, x)$	from 2
4	$\forall x\ Likes(x, a)$	from 1, 3
5	$Likes(b, a)$	from 4
6	$\exists x\ Likes(b, x)$	from 5
7	$\forall x\ Likes(x, b)$	from 1, 6

Recall that we referred above to natural-deduction proof calculi, in which each connective and quantifier is associated with a pair of inference rules, one for introducing and one for eliminating. Were this calculus to be applied to the sequence immediately above, the rule of inference for eliminating the universal quantifier would sanction moving from

$$\forall x \forall y \left[(\exists z\ Likes(x, z)) \rightarrow Likes(y, x) \right]$$

to the following – where *a* is substituted for *x*:

$$\forall y \left[(\exists z \, Likes(a, z)) \rightarrow Likes(y, a) \right]$$

The reader is invited to see how other such rules can be used to construct a fully formal proof out of the sequence, with help from the works cited above.

There are languages for knowledge representation that fall between $\mathcal{L}_{PC}$ and $\mathcal{L}_{FOL}$, and for the most part these are the languages that anchor the brand of KR&R supporting the Semantic Web.[11] These languages are more expressive than $\mathcal{L}_{PC}$, the language of the propositional calculus, but less expressive than the language $\mathcal{L}_{FOL}$ of first-order logic. And these languages are associated with their own proof calculi. These languages are generally those associated with *description logics*.[12] We don't have the space needed for a full exposition of such logics, but fortunately they can in general be quickly characterized with reference to the ingredients that compose the propositional calculus and first-order logic. We shall refer to these logics as *point-k* logical systems. A particular system in the class will be named later when *k* is set to some natural number. The ins and outs of how the natural numbers work as indices is based on an idiosyncratic but straightforward table invented for ease of reference by Bringsjord to keep straight decidability theorems for the main logical systems standardly discussed in such contexts.[13] Using this table, here is what pins down point-two (i.e. point-*k* where *k* = 2) logic:

	Monadic relations	Dyadic relations	Triadic relations
None		•	•
One			
Unlimited	•		

The characterization of such systems is simple. To produce such logics, we simply begin by restricting the alphabet of $\mathcal{L}_{FOL}$ in various ways. As an example, we might insist that no triadic relation be allowed, that no dyadic relations be allowed, but that any number of monadic relations be allowed (as in the permutation shown in the table immediately above). The logical system with such a language (the language $\mathcal{L}_{P2}$) is *point-two logic*, or *monadic first-order logic*. Point-two logic will be otherwise just like FOL. A triadic relation is one that allows a relationship between three objects to be expressed. For example, the relation (*B*, let's say) of a natural number *n* being between two distinct natural numbers *m* and *j* is a triadic one; and here would be a truth regarding the natural numbers that involves this triadic relation:

$$\forall x \, \forall y \, \forall z \, (B(x, y, z) \rightarrow x \neq z)$$

This truth cannot be expressed in monadic FOL. Naturally, dyadic relations would range over two objects, and monadic relations over but one object.

Why would those logical systems between the propositional calculus and full first-order logic be so central to KR&R? The reason point-*k* logical systems are interesting and useful pertains to an aspect of logical systems that we have yet to discuss: namely, meta-properties of such systems. Important meta-properties of such systems include the meta-property of *decidability*. A logical system is decidable just in case there is an algorithm for determining whether or not a well-formed formula in the language in question is a theorem. Of course, assuming the the Church–Turing thesis is true, the existence of such an algorithm guarantees that there is a computer programme that can determine, given as input a $\varphi \in \mathcal{L}_{p2}$, whether φ is a theorem.[14] While the propositional calculus is decidable, the predicate calculus is not. However, point-two logic is decidable. From the standpoint of KR&R, this is thought by many to be quite desirable. The reason is clear, namely that queries against knowledge bases populated by formulas expressed in $\mathcal{L}_{p2}$ can always (eventually) be answered, that is, where Φ is such a knowledge base, queries of the form

$$\Phi \vdash \varphi \, ?$$

can, given enough time and working memory, always be answered by a standard computing machine.

A fact that beginner students of KR&R and logic often find quite surprising is that the moment even one dyadic relation is allowed into a logic otherwise like FOL, that logic becomes undecidable; the proofs are actually quite simple. However, if one allows another dimension of parameterization into the picture, namely the number of quantifiers allowed in formulas, one can allow an expansion on the relation side and yet still preserve decidability, as long as *k* is quite small. We must leave such details aside.

The final point that must be made in this section is that there are many, many (actually, an infinite number of) logical systems more expressive than first-order logic. We mention just two examples. The first is in the space of extensional logics, the second in the space of intensional logics.

The first example is *second-order logic* (SOL), which allows quantification over functions and relations, a phenomenon that routinely occurs in natural language. The formal language in question, $\mathcal{L}_{SOL}$, includes variables for functions and relations. For instance, it seems quite plausible not only that if John is the very same thing as the father of Bill, John and the father of Bill either both have or both lack the property of being obese, but more generally that these two entities are such that every relation is one they share or lack. The general principle operative would be that two things are one and the same just in case every attribute is one they either share or lack; this principle is known as Leibniz's Law. In SOL we can formalize this law as:

$$(LL) \; \forall x \forall y \, (x = y \leftrightarrow \forall X \, (Xx \leftrightarrow Xy))$$

Note that in (LL) the variable X ranges over relations, whereas x and y range over individual objects in the domain. Humans find it easy enough to discuss scenarios in which attributes themselves have properties, but we leave aside *third-order logic* and beyond.

Our second example is a simple *epistemic logic*, in which, to the propositional calculus, we add an operator **K** for 'knows', which allows us to represent such propositions as that Albert knows that Brian knows that p is the case, as follows,

$$\mathbf{K}_a \mathbf{K}_b p$$

from which we can deduce, using an axiom that is standard in such logics (namely, that if an agent knows φ, φ is true), that in fact Brian knows that p. A recent exploration of the applicability KR&R is based on advanced epistemic logic.[15]

We have now reached the point at which we can discuss the dividing line between philosophy-less and philosophy-infused KR&R.

A Proposed Dividing Line Between Philosophy-less KR&R and Philosophy-powered KR&R

The idea for a dividing line is really quite straightforward: KR&R can be productively pursued, and KBSs built, in the complete absence of philosophy – but only as long as the information represented and reasoned over is not in the realm of the formal sciences, nor in the realm of everyday sophisticated human socio-cognition. On the other hand, philosophy will need to be part and parcel of KR&R when that which is to be represented and reasoned over involves these realms. We can put this position in the form of a claim that makes reference to the expressiveness of formal languages of the sort canvassed above:

Claim $\mathcal{C}$ (Regarding the Relationship Between Philosophy and KR&R):

> KR&R that represents propositional content in formulas of a formal language less expressive than that used in full first-order logic ($\mathcal{L}_{FOL}$) is unable to represent and reason over propositions containing concepts routinely used in the formal sciences, and in everyday human socio-cognition, and as such, such KR&R will have no need for the field of philosophy. Moreover, to engineer KBSs able to represent and reason over the more demanding phenomena in these domains will require a contribution from philosophy, and will specifically require:
> 1. formulas in $\mathcal{L}_{FOL}$ that, once rewritten so that all quantifiers appear in a leftmost sequence in such formulas (i.e. once rewritten in *prenex normal form*[16]), are irreducibly populated by at least five non-vacuous quantifiers, must be allowed; and
> 2. formulas in formal languages that are more expressive than $\mathcal{L}_{FOL}$ must be allowed.

We turn now to concretizing this claim by discussing some challenges KR&R can meet only if both philosophy and the associated languages are employed.

The Need for Philosophy

If C is true, then it should be easy enough to see the need for philosophy and the associated representation and reasoning schemes by considering some examples from the relevant domains, and we turn to such consideration now. We first look at an example from the formal sciences, and then one from socio-cognition. In both cases, the KR&R in question has been pursued, and is in fact currently still underway, in our own laboratory.

KR&R and the Formal Sciences

KR&R allows for, indeed in large measure exists to enable, the issuing of queries against knowledge-bases. As such, there is clearly a vantage point from which to see the applicability of KR&R within the formal sciences, that is, within such fields as formal logic, game theory, probability theory, the various subfields of mathematics (e.g. number theory and topology), and so on. In fact, the part of formal logic known as *mathematical logic*, aptly and (given our present purposes) conveniently is sometimes also called 'meta-mathematics'. Kleene[17] explicitly provides this vantage point, as we now explain, in brief. After this presentation, we explain why the above conjecture's claim about the formal sciences seems to be quite plausible.

The first step is to view activity in the formal sciences from the standpoint of *theories*. By 'theory' here is meant something purely formal, not anything like, say, the 'theory' of evolution, which is usually disturbingly informal.[18] In the formal sense, a theory $\mathcal{T}_\Phi$ is a set of formulas deducible from a set of axioms Φ; more precisely, given Φ, the corresponding theory $\mathcal{T}_\Phi$ is

$$\{\varphi \in \mathcal{L} \colon \Phi \vdash \varphi\}$$

We say in this case that Φ are the axioms *of* the theory. Note that there is a background formal language $\mathcal{L}$ from which the relevant formulas are drawn.

But why might it be reasonable to regard *all* research in the formal sciences to revolve around theories? We don't have the space to fully articulate and defend this view, and hence rest content to convey the basic idea, which is straightforward. That idea is this: work in a formal science S can be idealized as the attempt to ascertain whether or not propositions of interest follow deductively from a core set of axioms for S; that is, whether these propositions of interest are indeed part of the theory that arises from the axioms for S. In this scheme, probability theory, game theory, mathematics, and so on each consists in the attempt to increasingly pin down the theory determined by the core axioms in question.

For an example simple enough to present here, let's consider a fragment of mathematics, namely, elementary arithmetic. Specifically, we consider the theory of arithmetic known as 'Peano Arithmetic', or simply as **PA**.[19] The axioms of **PA** are the sentences 1–6 (where s is the successor function, and the other symbols are interpreted in keeping with grade-school arithmetic; e.g. × is ordinary multiplication), and any sentence in the universal closure of the Induction Schema, 7.[20]

1. $\forall x\,(0 \neq s(x))$ 4. $\forall x\,\forall y\,(x + s(y) = s(x + y))$
2. $\forall x\,\forall y\,(s(x) = s(y) \rightarrow x = y)$ 5. $\forall x\,(x \times 0 = 0)$
3. $\forall x\,(x + 0 = x)$ 6. $\forall x\,\forall y\,(x \times s(y) = (x \times y) + x)$
7. $[\varphi(0) \wedge \forall x(\varphi(x) \rightarrow \varphi(s(x)))] \rightarrow \forall x\varphi(x)$

Given this machinery, the part of mathematics known as arithmetic can be viewed as an attempt to increasingly pin down $\mathcal{T}_{PA}$. And now it should be clear, in turn, why KR&R can be regarded as having direct applicability. One reason is that **PA** can be thought of as a knowledge-base, and the attempt to make more and more progress figuring out what is and isn't in the theory $\mathcal{T}_{PA}$ can be viewed as the attempt to ascertain whether or not, for various formulae generable from the language of arithmetic, say φ,

$$\mathbf{PA} \vdash \varphi$$

We could thus view 'progress in the field of arithmetic' to be the answer to questions such as whether or not it's true that 29 plus 0 equals 29, and whether or not it's true that 3,000 times 0 equals 0, and so on.

Why, in light of the foregoing and other material, is the claim $\mathcal{C}$ plausible? The answer, put non-technically, is quite straightforward; and comes in three parts, to wit:

- Courtesy of Gödel's first incompleteness theorem, we know that there are truths about arithmetic that cannot be proved from **PA**.[21]
- Thanks to additional formal work, we know that some of these truths can nonetheless be proved.[22] Let's call this set $\hat{G}$.
- The general nature of the representation and reasoning needed to establish the truths in $\hat{G}$ is an open question in KR&R and philosophy, but it is clear that full first-order logic is required (for the simple reason that formulas in $\hat{G}$ require $\mathcal{L}_{FOL}$ to be expressed).

So here we have an ongoing investigation in the intersection of the formal sciences and KR&R that both intersects with philosophy, and is consistent with claim $\mathcal{C}$.

KR&R and Socio-Cognition

We now give our second example of $\mathcal{C}$ 'in action' by considering a specific concept in the sphere of socio-cognition. The concept is mendacity. We shall show that careful KR&R in this area necessitates both philosophy, and very expressive formal languages for knowledge representation.

Mendacity

We introduce the topic of mendacity in connection with KR&R by asking you to consider a confessedly idealized scenario.[23] The scenario involves a superficial and implausible concept of lying, but as a warm-up to the genuine article, we indicate how the machinery of unsophisticated KR&R can be brought to bear to provide a solution to the scenario. Afterward, we present a philosophically inspired, plausible definition of lying and demonstrate how a sophisticated, philosophically informed, KR&R system can be used to distinguish lies and liars from honesty and the honest. Without further prelude, we ask you to consider the following scenario: you have been sent to the war-torn and faction-plagued planet of Raq. Your mission is to broker peace between the warring Larpal and Tarsal factions. In a pre-trip briefing, you were informed that the Larpals are sending one delegate to the negotiations, and the Tarsals are sending a pair. You were also warned that Larpals are liars, i.e. whatever they say is false, while Tarsals are not, i.e. whatever they say is true. Upon arrival, you are met by the three alien delegates. Suddenly, you realize that though the aliens know whom among them are Larpals, and whom are Tarsals, you do not. So, you ask the first alien, 'To which faction do you belong?' In response, the first alien murmurs something you can't decipher. Seeing your look of puzzlement, the second alien says to you, · 'It said that it was a Larpal'. Then, with a cautionary wave of an appendage and an accusatory glance at the second alien, the third alien says to you 'That was a lie!' Whom among the three aliens can you trust?

Resolution of the Larpals and Tarsals scenario, at least in its present form, requires no more sophistication than $\mathcal{L}_{PC}$ and reasoning therewith. The scenario is recast into $\mathcal{L}_{PC}$ by, say, representing the three aliens with three constants, their factional membership (Larpal or Tarsal) as mutually exclusive properties, and their assertions as conditional formulas. In Figure 10.1, we show the scenario thus represented, and automatically solved, in Athena,[24] a KR&R system based on multi-sorted, first-order logic, and integrated with both the Vampire theorem prover and the Paradox model finder. The solution to this scenario, expressed in English, is as follows: the second alien is either a Larpal or a Tarsal. If it is a Tarsal, then truly the first alien said that it was a Larpal. Yet, if the first alien said that it was a Larpal, then it told the truth because a Tarsal would not lie and say it was a Larpal, but in so telling the truth, the first alien has distinguished itself as a

Tarsal – a contradiction! Ergo, the second alien cannot be a Tarsal; it is a Larpal. Therefore, the first and third aliens are Tarsals, and thus trustworthy.

Though the Larpals and Tarsals scenario nicely illustrates unsophisticated KR&R in action, the fact of the matter is that the concept of lying used in the scenario is, as we have already indicated, simple-minded. In real life, the idea that liars' propositional claims are always materially false is, well, silly. We might reserve such phrases as *habitual liar* or *pathological liar* for such beings, but in the real world, even pathological liars sometimes assert true propositions, if only by accident. Likewise, it is utterly ūnrealistic to expect honest agents to be infallible, i.e. to expect that their assertions are always materially true, because honest agents, nevertheless, may state false propositions out of ignorance, or error in belief.

To say that an agent is a *liar* presupposes that one has at hand an account of what it is to lie – yet no such account was set out, let alone included in the knowledge-base constructed above for the Larpals and Tarsals scenario. Any reasonable account of lying must include not just what an agent does – the *actus reus* of lying – but also what the agent believes and intends. Mendacity and less egregious forms of deception are consummate only when an agent acts with the *mens rea* to deceive, i.e. when an agent acts intending others to hold beliefs

```
(domain Alien)  # there is a domain of Aliens.
(declare (A1 A2 A3) Alien)   # A1, A2, and A3 are Aliens.

# Larpal and Tarsal are properties of Aliens.
(declare (larpal tarsal) (-> (Alien) Boolean))

# each Alien is either a Larpal or a Tarsal, but not both.
(assert (and (or (larpal A1) (tarsal A1))
    (not (and (larpal A1) (tarsal A1)))))
(assert (and (or (larpal A2) (tarsal A2))
    (not (and (larpal A2) (tarsal A2)))))
(assert (and (or (larpal A3) (tarsal A3))
        (not (and (larpal A3) (tarsal A3)))))

# among A1, A2 & A3 are one Larpal and two Tarsals.
(assert (iff (larpal A1) (and (tarsal A2) (tarsal A3))))
(assert (iff (larpal A2) (and (tarsal A1) (tarsal A3))))
(assert (iff (larpal A3) (and (tarsal A1) (tarsal A2))))

# if A3 is a Larpal, then A2 is a Tarsal...
(assert (if (larpal A3) (tarsal A2)))

# and if A2 is a Tarsal, then A1 said that it is a Larpal.
(assert (if (tarsal A2) (larpal A1)))

# but if A1 said that it is a Larpal, then it is a Tarsal!
(assert (if (larpal A1) (tarsal A1)))

Athena transcript:

>(load-file "larpals-and-tarsals.ath")
...
>(!prove (and (tarsal A1) (larpal A2) (tarsal A3)))
Theorem: (and (tarsal A1) (larpal A2) (tarsal A3))
```

Figure 10.1: Larpals and Tarsals scenario resolved in Athena.

that are contrary to what the agent believes to be true. To illustrate, assume that Amy is asked in geography class to name the state capital of California. Amy, erroneously believing that Los Angeles is the capital of California, answers with Los Angeles. Though Amy's answer is materially false, we would not ordinarily accuse Amy of lying because she has answered faithfully according to her belief – her statement was truthfully made, though it was not factually true. However, had Amy known that Sacramento is the capital of California, but answered Los Angeles intending to give a false impression of at least her own mind, then, indeed, she would have been lying. Now assume that Bob is helping Carl, a fugitive, flee from the police. The two agree that Carl should begin a new life in Canada, and then part ways. Later, when the police question Bob about Carl's whereabouts, Bob, intending to misdirect the police, tells them that Carl has gone to Mexico. Yet unbeknownst to Bob, Carl has changed his mind (and destination), moving to Mexico instead of Canada. Thus, Bob's statement to the police is materially true, though we would normally say that Bob lied because he believed that what he said was false, and said it intending to deceive – though factually true, his statement was falsely made. As these examples illustrate, the *mens rea* for lying and deception depends on the relationship between an agent's beliefs and the beliefs the agent intends for others.

Now, drawing upon philosophy, we set out a plausible definition of lying. We present this definition first informally, and then formally, using the language of a logical system, namely the socio-cognitive calculus (*SCC*). Once lying is thus defined, we explain, by revisiting the Larpals and Tarsals scenario, how a highly sophisticated KR&R system can prove, say, that an agent is a liar, or that one agent has lied to another.

Philosophy has a long tradition of contemplating the nature of mendacity and positing definitions thereof (a tradition going back at least to St Augustine). For exposition, we adopt Chisholm's account of lying – a seminal work in the study of mendacity and deception. Using *L* and *D* to represent correspondingly the speaker (i.e. the *liar*) and the hearer (i.e. the would-be *deceived*), we paraphrase below definitions of *lying* and the supporting act of *asserting*.

L lies to D = $_{df}$ There is a proposition *p* such that (i) either *L* believes that *p* is not true or *L* believes that *p* is false and (ii) *L* asserts *p* to *D*.[25]

L asserts p to D = $_{df}$ *L* states *p* to *D* and does so under conditions which, he believes, justify *D* in believing that he, *L*, accepts *p*.[26]

Chisholm and Feehan's conception of lying is that of promise breaking.[27] Assertions, unlike non-solemn (e.g. ironic, humorous or playful) statements, proffer an implicit social concord: one that offers to reveal to the hearer the mind of the speaker. In truthful, forthright communication, the speaker fulfills the promise and obligation of this concord. In lying, the speaker proffers the concord in bad faith: the speaker does not intend to, and does not, fulfill the obligation to reveal his/her true mind, but instead reveals a pretense of belief. In this way, lying 'is essentially a breach of faith'.

The above is, of course, a highly condensed presentation of work, and there are various nuanced philosophical facets to it.[28] Yet, even in condensed form, it is evident that the concepts of *lying* and *asserting* depend on agents' temporally coupled beliefs and actions. Thus, formal definition of these concepts requires the use of highly expressive languages for KR&R: ones that can represent, and allow reasoning over, the beliefs and actions of agents through time.

To formally define lying and asserting, we employ the *socio-cognitive calculus* (*SCC*). The *SCC*[29] is a KR&R system for representing, and reasoning over, events and causation, and perceptual, doxastic and epistemic states (it integrates ideas from the event calculus and multi-agent epistemic logic). The *SCC* is an extension to the Athena system, providing, among other things, operators for perception, belief, knowledge and common knowledge. The signature and grammar of the *SCC* is shown following. Since some readers may not be familiar with the concept of a signature, we note that it is simply a set of announcements about the categories of objects that will be involved, and about the functions that will be used to talk about these objects. Thus it will be noted that in Figure 10.2, the signature in question includes the specific announcements that one category includes agents, and that *happens* is a function that maps a pair composed of an *event* and a *moment*, and returns true or false (depending upon whether the event does or does not occur at the moment in question).

Sorts	$S ::=$	Object | Agent | ActionType | Action | Event | Fluent | Modern | Boolean
		action: Agent × ActionType → Action
		initially: Fluent → Boolean
		holds: Fluent × Moment → Boolean
		happens: Event × Moment → Boolean
Functions	$f ::=$	*clipped*: Moment × Fluent × Moment → Boolean
		initiates: Event × Fluent × Moment → Boolean
		terminates: Event × Fluent × Moment → Boolean
		prior: Moment × Moment → Boolean
Terms	$t ::=$	$x: S \mid c \cdot S f(t_1 \ldots t_n)$
		t: Boolean $\mid \neg P \mid P \wedge Q \mid P \rightarrow Q \mid P \leftrightarrow Q \mid \forall_{x:S} P \mid$
Propositions	$P ::=$	$\exists_{x:S} P \mid \mathbf{S}(a, P) \mid \mathbf{K}(a,P) \mid \mathbf{B}(a,P) \mid \mathbf{C}(P)$

Figure 10.2: Example of the socio-cognitive calculus.

Reasoning in the $\mathcal{SCC}$ is realized via natural-deduction-style inference rules. For instance, R_2 shows that knowledge entails belief; R_3 infers from 'P is common knowledge' that, for any agents a_1, a_2 and a_3, 'a_1 knows that a_2 knows that a_3 knows that P'. And R_4 guarantees the veracity of knowledge; that is, if an agent 'knows that P', then P is, in fact, the case.

$$\frac{}{\mathbf{C}(\mathbf{K}(a,P) \to \mathbf{B}(a,P))} [R_2] \qquad \frac{\mathbf{C}(P)}{\mathbf{K}(a_1, \mathbf{K}(a_2, \mathbf{K}(a_3, P)))} [R_3] \qquad \frac{\mathbf{K}(a,P)}{P} [R_4]$$

In the $\mathcal{SCC}$, agent actions are modelled as types of events. We model lying, asserting, and stating propositions as types of actions that an agent may perform. These action types are denoted by the functions *lies*, *asserts* and *states*. The argument to such action types are conceived of as reified propositions, specifically fluents. Thus, the formula happens(*action*(l, *states*(p, d)), m) is read, 'it happens at moment m that agent l states (reified) proposition p to agent d'. For convenience, we model that an agent is a liar by using the property *liar*. The signature for these additions to the $\mathcal{SCC}$ is as follows (see Figure 10.3):

Functions $\quad f ::= \quad$
$$states: \text{Fluent} \times \text{Agent} \to \text{ActionType}$$
$$asserts: \text{Fluent} \times \text{Agent} \to \text{ActionType}$$
$$lies: \text{Fluent} \times \text{Agent} \to \text{ActionType}$$
$$liar: \text{Agent} \to \text{Boolean}$$

Figure 10.3: Signature in the socio-cognitive calculus.

The definitions of *liar*, *lies* and *asserts* are stipulated as common knowledge by axioms (1)–(3).

$$\mathbf{C}\left(\forall_l \; liar\,(l) \leftrightarrow \exists_{d,p,m} \; happens \; (action \; (l, lies \; (p,d)), m)\right) \tag{1}$$

$$\mathbf{C}\left(\begin{array}{l} \forall_{l,d,p,m} \; happens \; (action \; (l, lies \; (p,d)), m) \leftrightarrow \\ \qquad \mathbf{B}(l, \neg \, holds \; (p,m) \; \wedge \\ happens \; (action \; (l, asserts \; (p,d)), m)) \end{array}\right) \tag{2}$$

$$\mathbf{C}\left(\begin{array}{l} \forall_{l,d,p,m} \; happens \; (action \; (l, asserts \; (p,d)), m) \leftrightarrow \\ \qquad \left(\begin{array}{l} happens \; (action \; (l, states \; (p,d)), m) \; \wedge \\ \mathbf{B}(l, \mathbf{B}(d, happens \; (action \; (l, states \; (p,d)), m) \to \\ \qquad \mathbf{B}(l, holds \; (p,m)))) \end{array}\right) \end{array}\right) \tag{3}$$

Now that we have in hand a formal account for lying, we can re-examine the scenario posed earlier. Assuming that Larpals conform to the plausible definition of lying – not that every statement they make is false, but rather that their assertions, at times, misrepresent their beliefs – and that Tarsals conform to a counterpart notion of honesty – that their assertions faithfully reflect their beliefs, which are, however, still fallible – can one determine which aliens are trustworthy Tarsals?

In order to represent the Larpals and Tarsals scenario, we further extend the *SCC* signature with the functions *alien*, *larpal* and *tarsal* with a number of constants (see Figure 10.4):

$$
\text{Functions} \quad f ::= \quad
\begin{aligned}
&alien : \text{Agent} \rightarrow \text{Boolean} \\
&larpal : \text{Agent} \rightarrow \text{Boolean} \\[1em]
&tarsal : \text{Agent} \rightarrow \text{Boolean}
\end{aligned}
$$

$$
\text{Constants} \quad c ::= \quad
\begin{aligned}
&H, A_1, A_2, A_3 : \text{Agent} \\
&p_1, p_2, p_3 : \text{Fluent} \\
&m_1, m_2, m_3 : \text{Moment}
\end{aligned}
$$

Figure 10.4: Signature extension.

The constants A_1, A_2 and A_3 denote the three distinct aliens, and the constant H denotes the human. From the scenario we extract several pieces of common knowledge: (i) A_1, A_2 and A_3 are aliens, while H is not; (ii) Larpals are liars, and Tarsals are not; (iii) every alien is a Larpal or a Tarsal; and (iv) aliens recognize whether other aliens are Larpals or Tarsals. This common knowledge is represented by axioms (4)–(7):

$$\mathbf{C}(alien\,(A_1) \land alien\,(A_2) \land alien\,(A_3) \land \neg\, alien\,(H)) \tag{4}$$

$$\mathbf{C}(\forall_a (larpal\,(a) \rightarrow liar\,(a) \;\land (tarsal\,(a) \rightarrow \neg\, liar\,(a)\,) \tag{5}$$

$$\mathbf{C}(\forall_a alien\,(a) \rightarrow (larpal\,(a) \lor tarsal\,(a))) \tag{6}$$

$$\mathbf{C}\left(\begin{aligned}
&\forall_{a_1,a_2}\,(alien(a_1) \land alien(a_2)\;\rightarrow \\
&((tarsal(a_2) \rightarrow \mathbf{K}(a_1, tarsal(a_2)))\,\land \\
&(larpal(a_2) \rightarrow \mathbf{K}(a_1, larpal(a_2)))\quad\;
\end{aligned}\right) \tag{7}$$

In addition to the above common knowledge, H knows that there are exactly two Tarsals and one Larpal in the delegation. This knowledge is represented by axiom (8).

$$\mathbf{K}\begin{pmatrix} (larpal\ (A_1) \leftrightarrow (tarsal\ (A_2) \wedge tarsal\ (A_3))) \wedge \\ H, (larpal\ (A_2) \leftrightarrow (tarsal\ (A_1) \wedge tarsal\ (A_3))) \wedge \\ (larpal\ (A_3) \leftrightarrow (tarsal\ (A_1) \wedge tarsal\ (A_2))) \end{pmatrix} \tag{8}$$

The specific interactions in the scenario at hand brings about several more axioms. The first alien's utterance was unclear, but it is common knowledge that, at moment m_1, it asserted something to H; that something is denoted by the reified proposition p_1. Similarly, it is common knowledge that, at moment m_1, the second alien asserted p_2 to H. Furthermore, it is common knowledge that p_2 is materially true if and only if the first alien declared itself a Larpal. Finally, it is common knowledge that, at moment m_3, the third alien asserted p_3 to H, and that p_3 is materially true if and only if the second alien's assertion to H was a lie. These actions, and the truth conditions of the various assertions, are represented by axioms (9)–(13).[30]

$$\mathbf{C}(happens(action(A_1, asserts(p_1, H)), m_1)) \tag{9}$$

$$\mathbf{C}(happens(action(A_2, asserts(p_2, H)), m_1)) \tag{10}$$

$$\mathbf{C}(holds(p_2, m_2) \leftrightarrow (holds(p_1, m_1) \leftrightarrow larpal\ (A_1))) \tag{11}$$

$$\mathbf{C}(happens(action(A_3, asserts(p_3, H)), m_3)) \tag{12}$$

$$\mathbf{C}(holds(p_3, m_3) \leftrightarrow happens(action(A_2, lies(p_2 H)), m_2)) \tag{13}$$

With the Larpals and Tarsals scenario now formalized in *SCC*, we proceed to sketch how H, given sufficient contemplation, can know that A_1 and A_3 are Tarsals, and that A_2 is a Larpal. Our sketch consists of three parts: (i) we indicate how H can know that if A_2 is a Tarsal, then A_3 is a Larpal; (ii) we indicate how H can know that if A_1 is a Tarsal, then A_2 is a Larpal; (iii) we indicate how H can know, based on these two conditionals, that A_1 and A_3 are Tarsals, and that A_2 is a Larpal. In the prose elaboration of the three parts, the reasoning is described from H's perspective.

First, suppose (H reasons to itself) that A_2 is a Tarsal. Faction membership is apparent to aliens, and so A_3 also knows that A_2 is a Tarsal. A_3 also knows that Tarsals are not liars, and, more specifically, that it does not happen that A_2 lies to H. Therefore, A_3 knows that the proposition that it asserted, p_3, does not hold, i.e. it is materially false. Since A_3 knows this, it also believes this. Hence, A_3 was lying when it made its assertion, so it must be a liar, and so not a Tarsal, and thus a Larpal. In this way, H reasons to itself that if A_2 is a Tarsal, then A_3 is a Larpal. Here, expressed in the aforementioned 'streamlined' format for describing a proof, is an abbreviated proof that mirrors this description of H's reasoning:

1   $\mathbf{K}(H, tarsal(A_2))$ assumption

2   $\mathbf{K}(H, \mathbf{K}(A_3, alien(A_2)))$ by Axiom (4)

3   $\mathbf{K}(H, \mathbf{K}(A_3, tarsal(A_2)))$ by Axiom (7) and step 1

4   $\mathbf{K}(H, \mathbf{K}(A_3, \neg liar(A_2)))$ by Axiom (5) and step 3

$$5 \quad \mathbf{K}\left(H, \mathbf{K}\left(\begin{array}{l} A_3, \\ \neg\, happens\left(action\left(\begin{array}{l} A_2, \\ lies\,(p_2, H) \end{array} \right), m_2 \right) \end{array} \right)\right) \qquad \text{by Axiom (1) and step 4}$$

6   $\mathbf{K}(H, \mathbf{K}(A_3, \neg holds(p_3, m_3)))$ by Axiom (13) and step 5

7   $\mathbf{K}(H, \mathbf{B}(A_3, \neg holds(p_3, m_3)))$ by step 6 and R_2

8   $\mathbf{K}(H, happens(action(A_3, lies(p_3, H))m_3))$ by Axioms (2) and (12) and step 7

9   $\mathbf{K}(H, liar(A_3))$ by Axiom (1) and step 8

10   $\mathbf{K}(H, \neg tarsal(A_3))$ by Axiom (5) and step 9

11   $\mathbf{K}(H, larpal(A_3))$ by Axiom (6) and step 10

Next, suppose (H reasons to itself) that A_1 is a Tarsal. Faction membership is apparent to aliens, and so A_2 knows that A_1 is a Tarsal. A_2 also knows that Tarsals are not liars, and, more specifically, that it does not happen that A_1 lies to H. Since A_2 knows that A_1 asserted p_1 to H and A_1 does not lie, A_2 knows that it is not the case that A_1 believes that p_1 does not hold. Yet, A_2 also knows A_1 knows (and thus believes) that A_1 is not a Larpal. Therefore, A_2 knows that it is impossible for A_1 to have asserted that it is a Larpal, for if it did, then it would be a liar. That is to say, A_2 knows that its own assertion p_2 does not hold, i.e. it is materially false. Thus, A_2 lies in asserting p_2. In this way, H reasons to itself that if A_1 is a Tarsal, then A_2 is a Larpal.

Last, were it the case (H reasons to itself) that A_1 is a Larpal, then A_2 and A_3 would be Tarsals, because there is only one Larpal among the three. Yet, if A_2 is a Tarsal, then A_3 is a Larpal, which contradicts A_3 being a Tarsal. Hence, A_1 is not a Larpal, thus A_1 is a Tarsal. Since A_1 is a Tarsal, A_2 is a Larpal, and then A_3 is, like A_1, a Tarsal. Finally, in this way, H reasons to itself that A_1 and A_3 are trustworthy Tarsals, and A_2 is the dishonest Larpal.

Note that in the final part, where H definitively determines which aliens are trustworthy, H's reasoning depends on knowing that there are two Tarsals and one Larpal among the three aliens. Without such knowledge, it is impossible for H to decide who to trust. Alas, in the real world, such knowledge is not likely. Anyone, at least any human, may lie. Furthermore, nefarious plots (e.g. fraud, pyramid schemes, espionage, guerrilla tactics and terrorism) depend on lying and lesser deceptions. Machines may play a role in guarding e.g. free-market consumers, private citizens and sovereign states against such plots, but only if machines are able to comprehend philosophical concepts like mendacity and deception. In turn, KR&R systems cannot begin to grasp such concepts unless they embrace philosophy, and the formal sophistication that philosophy demands.

Brief Remark on Evil KBSs

In general, there seems to be no reason in principle why KR&R cannot be applied not only to socio-cognitive concepts like mendacity and deception, but also to even richer and more nuanced concepts that incontestably require philosophical analysis in order to be couched in terms precise enough to allow knowledge bases to hold queryable information about them. For instance, one could consider the possibility of engineering a KBS that is capable of betraying someone, or capable, in general, of being evil. It seems quite undeniable that no KR&R expert could engage in such engineering without both engaging philosophy and making use of highly expressive logics.

Engineering for the former case, which did indeed explicitly involve both philosophy and highly expressive formal languages, has already been carried out by Bringsjord.[31] In this work, philosophical analysis was used to gradually craft a definition of the concept of one agent betraying another. This definition was consistent with C.

What about evil? Here the investigation is still in its early stages.[32] The basic process, though, is the same as what we showed in action in connection with mendacity: philosophy is used to build the definition of evil; the definition is formalized in some logical system; knowledge bases describing evil agents are populated; and queries against such knowledge bases are issued and answered, which gives rise to the relevant knowledge-based systems. The interesting thing about this KR&R work is that if, as some have claimed,[33] a truly evil agent is one that harbours outright contradictions in what he or she believes, logical systems able to allow the representation of contradictory information, and the unproblematic reasoning over that information, would be necessary. Such logical systems are highly expressive and would be C-confirming. These systems are known as *paraconsistent logics*.[34]

'Visual' KR&R and the Future

Heretofore, when representing propositional content, the field of KR&R has been exclusively linguistic in nature.[35] This is consistent with the fact that, to this point in the present essay, all formal languages used for the representation of propositional content have been exclusively linguistic: Well-formed formulas generable by the alphabets and grammars of these languages are invariably strings of characters, and these strings in no way 'directly resemble' that which they are intended to denote. For example, when we spoke earlier of liars and truthtellers, and used names to refer to them in our case studies of mendacity, we specifically used the constant 'A_1' to refer to one of the aliens. Had we felt like doing so, we could just as easily have used instead the constant 'A99', or 'a1' or 'A-23', and so on, *ad indefinitum*. In contrast, a diagrammatic representation of the alien in question would bear a resemblance relation to him, and even slight changes in the diagram could prevent it from denoting the alien. As the philosopher Peirce put it, 'a diagram is naturally analogous to the thing represented'.[36]

Despite the fact that KR&R has traditionally left aside pictorial representation schemes, there can be no disputing the fact that human reasoning is powerful in no small part because it is often *diagrammatic* in nature. Ironically, while KR&R, as elegantly explained by Glymour,[37] has been purely linguistic since the first formal language for KR&R was introduced by Aristotle (namely, the theory of the syllogism), Aristotle, along with his linguistic-oriented successors in the modern era (e.g. Boole and Frege), sought to explain how the highly visual activity of the great Euclid could be formalized via some logical system. It is plausible to hold, as we do, that substantive parts of this long-sought explanation began to arrive on the scene courtesy of seminal work carried out by a pair of oft-collaborating logician/philosophers: Jon Barwise and John Etchemendy. Since the space we have to discuss diagrammatic KR&R is quite limited, we shall briefly explain, by way of a problem posed by this pair, how a hybrid diagrammatic/linguistic formal language in the so-called Vivid family of such languages can be used to solve this problem. The problem is a seating puzzle given in Barwise and Etchemendy.[38]

Here is the seating puzzle, which has become something of a classic: Five people – *A, B, C, D, E* – are to seated in a row of seats, under the following three constraints.

C1 *A* and *C* must flank *E*.

C2 *C* must be closer to the middle seat than *B*.

C3 *B* and *D* must be seated next to each other.

Now, three problems are to be solved:

P1 Prove that *E* cannot be in the middle, or on either end.

P2 Can it be determined who must be sitting in the middle seat?

P3 Can it be determined who is to be seated on the two ends?

The class of relevant diagrams in this case can conveniently be viewed as a quintuple, each member of which is either one of the five people, or the question mark. For example, here is a diagram:

AECBD

Note that this diagram satisfies all three constraints. As another example, the diagram

??A??

is one in which *A* is seated in the middle chair.

We are now in position to consider a proof that constitutes a solution to the puzzle. This proof will be *heterogeneous*: it will make use of propositional content expressed in traditional form (i.e. in the form of formulas from the kind of formal languages presented and employed above), *and* it will make use of such content expressed in the form of diagrams. Since none of the formal languages seen above (e.g. $\mathscr{L}_{PC}$, $\mathscr{L}_{FOL}$, etc.) allow for diagrams as well-formed expressions, this proof cannot be based in the logical systems visited above. Here is the proof:

Proof. Given that E must be between A and C, there are six diagrams to consider, namely:

AEC?? (1)	?*AEC*? (2)	??*AEC* (3)
CEA?? (4)	?*CEA*? (5)	?? *CEA* (6)

However, only (1) and (6) are consistent with the other two constraints. P1 is therefore accomplished. Since in both of these diagrams C is in the middle seat, P2 is answered in the affirmative. As to P3, the answer is 'No', since any end could have one of A, B or D. QED

The diagrammatic representations seen in this seating puzzle, and in the solution thereof, are frequently used by human reasoners, but have hitherto not been part and parcel of KR&R. Yet, in philosophy, there is a very strong tradition of not only recognizing that such representations are often used, but also of making their use precise in systems that go beyond the purely linguistic. We do not have the space to present and discuss such systems here. We direct the reader to the Vivid system for details on how the seating puzzle, as well as much more complicated representation and reasoning of a visual sort, can be made precise and mechanized.

We conclude this section with a brief remark about the historical context. Logic grew out of philosophy; computer science, and specifically AI, in turn, grew out of logic. This progression is nicely chronicled by Glymour. But we are now in a *new* progression, one driven by philosophy and logic as midwives, and is gradually expanding KR&R into the visual realm.

Conclusion

We have set out and defended the view that if KR&R is to reach into the realms of mathematics and socio-cognition, then philosophy must become a genuine partner in the enterprise. While we have mentioned a number of phenomena in these realms, we have shown this view in action through a particular emphasis on a concept – mendacity – that by its very nature involves social cognition. We predict that as KR&R expands and matures in the future – if for no other reason than that it should allow humans to work collaboratively with intelligent machines having direct and immediate access to electronic propositional content about social, cognitive, mathematical and visual matters – philosophy and philosophers will be consulted to an increasingly high degree. If such consultation does not come to pass, and the conjecture above is correct, it follows that KR&R will be limited to propositional content that is but a tiny fragment of what is known by human beings; and it follows in turn from that that intelligent machines, relative to human minds, will, knowledge-wise, remain exceedingly primitive by comparison.

11 ON FRAMES AND THEORY-ELEMENTS OF STRUCTURALISM

Holger Andreas

Introduction

There are quite a few success stories illustrating philosophy's relevance to information science. One can cite, for example, Leibniz's work on a *characteristica universalis* and a corresponding *calculus ratiocinator* through which he aspired to reduce reasoning to calculating. It goes without saying that formal logic initiated research on decidability and computational complexity. But even beyond the realm of formal logic, philosophy has served as a source of inspiration for developments in information and computer science. At the end of the twentieth century, formal ontology emerged from a quest for a semantic foundation of information systems having a higher reusability than systems being available at the time. A success story that is less well documented is the advent of *frame systems* in computer science. Minsky is credited with having laid out the foundational ideas of such systems.[1] There, the logic programming approach to knowledge representation is criticized by arguing that one should be more careful about the way human beings recognize objects and situations. Notably, the paper draws heavily on the writings of Kuhn and the *Gestalt*-theorists.

It is not our intent, however, to document the traces of the frame idea in the works of philosophers. What follows is, rather, an exposition of a methodology for representing scientific knowledge that is essentially frame-like. This methodology is labelled as *structuralist theory of science* or, in short, as *structuralism*.[2] The frame-like character of its basic meta-theoretical concepts makes structuralism likely to be useful in knowledge representation. After a brief introduction to frames in the next section, the basic concepts of structuralism are outlined in the third section (pp. 122–6). We deal with structuralist knowledge representation in the final section (pp. 126–9).

Frames

What distinguishes the representation of putative knowledge by frames from representation by means of a logical first order theory? The latter kind of representation is characterized as flat in the sense that each piece of knowledge is represented independently of any other piece. The standard way to represent factual knowledge in the logic programming approach, which is based on a fragment of first order logic, is to use atomic sentences. Unless the interpretation of the language is constrained by so-called meaning postulates, any truth-value assignment to the atomic sentences that satisfies the semantic rules for logical connectives and quantifiers is considered admissible. This contrasts with frame systems, where the interpretation of the language is much more constrained and, because of this, also more structured.

Frames, essentially, have slots, i.e. properties that admit only certain *fillers* as values. Objects are represented as instances of individual frames, whereas concepts are represented as generic frames. Moreover, frames, usually, have subframes by means of which the specialization of a more general frame is represented. The relation of specialization is such that a subframe *inherits* the values of slots of the more general frame in a defeasible way. These ideas have found their way into *description logics* and *object oriented programming*.

Minsky himself envisioned frames to be related to one another in a domain-dependent way.[3] This means that there are relations between frames other than that of specialization. His primary idea was to model recognition of objects or situations by an interrelated system of frames such that it can be adopted for the purpose of automatization.

Basic Concepts of Structuralism

Frames in Structuralism

The structuralist representation scheme seems to qualify as a frame system in the very sense that this notion has been introduced by Minsky. Notably, there is profound resemblance between Minsky[4] and Sneed[5] regarding their general train of thought and the results established. Both Minsky and Sneed aim to overcome a certain kind of logical atomism without embracing a sort of holism that 'never materializes into a technical proposal', as stated by Minsky. Further, the notion of a paradigm in the sense of Kuhn[6] is being utilized in Minsky and Sneed, yet such utilization does not result in the rejection of formal or semi-formal means to analyse human conceptualization and scientific theories. Finally and most importantly, certain meta-theoretical concepts introduced in the structuralist framework simply behave like frames in the very sense this notion is understood in Minsky and subsequent work inspired by this paper. The latter feature will be exploited below.

The logical reconstruction of a theory in the structuralist framework is guided by the doctrine that a theory can be divided into several *theory-elements*. Roughly speaking, a theory-element is identified through a substantial claim, usually a law-like assertion or a conjunction of several law-like assertions. A scientific theory – in the non-technical sense of the term – is represented by a theory-net. Theory-nets, representing different theories, can be related to one another by so-called *links*.

Which types of frames are introduced in structuralism to model scientific theories? We may begin with the meta-theoretical concept of an *intended application*. This concept is to be taken in the literal sense that scientists have certain systems of empirical objects in mind to which, they think, a substantial claim of a particular theory applies. Such systems and their empirical properties are represented by certain set-theoretic entities whose typification is indicated in the next section. Importantly, it is systems of presumably interacting empirical objects, as opposed to single empirical objects, to which the substantial claims of a theory are modelled to apply. For example, in classical mechanics, the system of planets forming our solar system is considered, among other things, to be an intended application of Newton's second and third axiom.

What gives rise to the claim that the notion of an intended application is a frame-like concept? First, there is a strong typification of how intended applications of a theory-element are represented in structuralism by means of set-theoretic entities. The typification of entities and attributes is an essential feature of frame systems being inspired by Minsky's paper on frames. Second, an intended application originates from a human conceptualization of an ensemble of entities that is rather guided by *paradigms* than by formal definitions. Minsky stresses this aspect when he introduces the notion of a frame system.

While an instance of the class of intended applications represents only phenomenal properties of an empirical system, *potential models* do also contain information about the theoretical properties. Potential models are those entities of which it is sensible to ask whether the substantial claim of a theory-element is satisfied or not. If a potential model satisfies that claim, then it is called a model of a theory-element. In a certain sense, a theory-element comprises several interrelated frame-like concepts, namely the concept of an intended application, the concept of a potential model and the concept of a model. The distinction between phenomenal, or non-theoretical, properties and theoretical ones as well as the kind of slots associated with the frame-like concepts in structuralism will be explained in the next section.

Relations are introduced among frames in structuralism in the same way that they are introduced in the frame systems currently being used in software engineering. Here the relation of *specialization* among theory-elements is of particular importance. A theory-element that is a specialization of another theory-element, inherits certain properties of the more general theory-element and adds another substantial claim to the substantial claim of that theory-element. It is a conjecture expressed by Balzer and Moulines[7] that all theories employing some mathematical apparatus are tree-like, in the sense that there is one and only one theory-element which is not a specialization of any other theory-element, but to which every other element of the theory-net is in the relation of specialization. If there is such a theory-element, then it is called a fundamental one.

The relation of specialization is a relation among *frames* forming a theory-element. In addition, it was felt necessary to introduce concepts describing relations between instances of frames. So, the relational concept of a *constraint* accounts for relations among the intended application of one and the same theory-element, while the concept of a *link* accounts for relations among the intended applications of different theory-elements.

Set-Theoretic Entities in Structuralism

In structuralism, empirical systems are represented by set-theoretic entities of the following types:

$$\langle D_1, ..., D_k; A_1, ..., A_m; n_1, ..., n_p \rangle \tag{1}$$
$$\langle D_1, ..., D_k; A_1, ..., A_m; n_1, ..., n_p, t_1, ..., t_q \rangle \tag{2}$$

where $D_1,...,D_k$ are sets of empirical objects and $A_1,...,A_m$ sets of mathematical objects. The symbols $n_1,...,n_p$ designate non-theoretical relations; they are accordingly called non-theoretical terms. The symbols $t_1,...,t_q$ designate theoretical relations. In most cases, the theoretical and non-theoretical relations represent functions that map n-tuples of empirical objects onto mathematical objects. The domain and the range of these functions must be specifiable by means of set-theoretic operations on the sets $D_1,...,D_k$ and $A_1,...,A_m$. The sets D1,...,Dk comprise the empirical entities of a single system to which a theory is applied. In distinguishing between several sets of empirical objects within one single empirical system, one wishes to express type differences between empirical objects that are necessary to specify the domain and the range of the theoretical and the non-theoretical functions. Set-theoretic entities of the type $\langle D_1,...,D_k; A_1,...,A_m; R_1,...,R_n \rangle$ are called (set-theoretic) structures.

To model the application of a theory to a piece of the empirical world, the distinction between theoretical and non-theoretical relations is of crucial importance. Roughly, the idea is that a term t is theoretical with respect to a theory T, or T-theoretical, if its meaning cannot be explained without reference to T. More precisely, a term t is T-theoretical if there is no application of T in which the value of t can be determined independently of the axioms of T. Here, structures of type (2) are called theoretical structures, while structures of type (1) are called non-theoretical.

A theory-element is constructed as an ordered pair consisting of the theory-element's formal core and its set of intended applications:

$$T = \langle \mathbf{K}, \mathbf{I} \rangle \tag{3}$$

where **K** stands for the formal core and **I** for the set intended applications. **I** is a set of non-theoretical structures, i.e. structures of type (1). The formal core **K** of a theory-element can be thought of as a set-theoretic description of the pre-suppositions and the implications that the very act of applying a theory to the empirical world has. Most importantly, the formal core is designed to capture what is implied for the values of theoretical terms by the application of a theory, or a law of a theory, to a system of empirical entities. The formal core **K** of a theory-element **T** consists of several components:

$$\mathbf{K}(\mathbf{T}) = \langle \mathbf{M}_p(\mathbf{T}), \mathbf{M}(\mathbf{T}), \mathbf{M}_{pp}(\mathbf{T}), \mathbf{GC}(\mathbf{T}), \text{ and } \mathbf{GL}(\mathbf{T}) \rangle \quad (4)$$

$\mathbf{M}_p(\mathbf{T}), \mathbf{M}(\mathbf{T}), \mathbf{M}_{pp}(\mathbf{T}), \mathbf{GC}(\mathbf{T})$, and $\mathbf{GL}(\mathbf{T})$ are sets of structures or sets of sets of structures that are formally introduced by means of set-theoretic predicates. Our explanation of the components of **K**, the formal core of a theory-element, will be confined to indicating what function these components have in modelling the application of a theory:

- $\mathbf{M}_p(\mathbf{T})$, the set of potential models: a theoretical structure x is called a potential model of **T** if it satisfies certain frame conditions so that the question of whether x is model of the theory-element **T** is sensible.
- $\mathbf{M}(\mathbf{T})$, the set of models: a theoretical structure x is called a model of **T** if it satisfies the substantial claim of **T**, i.e. one or several law-like assertions that are associated with the theory-element **T**.
- $\mathbf{M}_{pp}(\mathbf{T})$, the set of partial potential models: a non-theoretical structure x is called a partial potential model of **T** if it can be augmented by theoretical relations so that the resulting theoretical structure is a potential model of **T**. It is assumed that the set of intended applications is a subset of the set of partial potential models.
- $\mathbf{GC}(\mathbf{T})$, the global constraint: if an empirical entity is a member of several empirical systems that are intended applications of **T**, then the assignment of values to the theoretical terms oftentimes is subject to certain restrictions. Such restrictions are represented by so-called constraints. The global constraint is the set-theoretic intersection of the single constraints of **T**.

- **GL(T)**, the global link: if an empirical entity is a member of several empirical systems that are intended applications of different theory-elements **T1** and **T2**, then the assignment of values to the theoretical terms oftentimes underlies certain restrictions. Such restrictions are represented by links. The global link represents the restrictions resulting from the set of single links.

Structuralist Knowledge Representation

Non-Automated Reasoning

Let us assume that a particular scientific theory *T*, being composed of several theory-elements, has been given a logical reconstruction in the framework of structuralism. What kind of information can be derived from such a reconstruction? First, there is some general information concerning the axioms, or substantial claims, of *T*:

- The substantial claims that define the potential models of the theory-elements of *T*;
- for every substantial claim, the concepts by which it is formulated;
- for every substantial claim, the type of entities about which this claim is made;
- the relations between the substantial claims in the form of relations between theory-elements;
- the relations between intended applications of one and the same theory-element in the form of constraints on the values of the theoretical relations.

The evaluation of scientific data seems to be more important than this general information. Regarding this task, two cases need to be distinguished. First, the scientific data encompass values of theoretical and non-theoretical relations. Second, these data encompass only values of non-theoretical relations. In the first case, we can determine whether the system of hypotheses is confirmed by the scientific data in a rather straightforward manner. To do this, we only need to examine whether substantial claims, which are encoded by the definition of the models of the theory-elements of the theory-net *T*, as well as the constraints among the intended applications of each theory-element are satisfied by the given data. These data are to be given as partial potential models of the theory-elements.

If the scientific data given encompass only non-theoretical relations, then the evaluation as to whether the system of hypotheses is confirmed by the data can proceed only on a certain condition, namely that the values of the non-theoretical relations determine the values of the theoretical relations uniquely. Arguably, this condition is met by a large number of scientific experiments. If, however, a unique determination of the theoretical relations is not guaranteed, the problem of whether the hypotheses of T are consistent with the data can only be solved by methods of automated reasoning. We will come back to this point below.

The case in which the hypotheses of one theory-net $T1$ are related to the hypotheses of another theory-net $T2$ adds more complexity. The technical term for such relation is 'link' (between theory-elements of different theory-nets). If links to other theory-nets are included in the reconstruction, then scientific data can be evaluated in the context of more than one theory-net. Since it is a major goal of ontological research to facilitate the interoperability of semantically heterogeneous data, this case deserves particular attention.

Automated Reasoning

In the preceding section, we considered a human being that elicits and reports meta-information about scientific data in a particular theoretical context on the basis of a structuralist reconstruction of that context. To automate the examination of scientific data, we need to transform the logical reconstruction of a theory-net into a machine readable format. For this, we will discuss two options: first, the transformation into a completely formalized theory; second, a transformation using semantic markup languages as the Resource Description Framework Schema (RDFS).

The structuralist representation scheme comes in a semi-formal fashion. The degree of formalization is, nevertheless, such that a complete formalization is not difficult to attain. Formulae in which expressions of naive set theory occur can be transformed into equivalent formulae using an axiomatization of set theory. Alternatively, one can transform set-theoretical expressions into those of higher order logic.[8]

As there exist theorem provers for higher order logic, there is one way to automate reasoning about a theory that is given in the structuralist framework. Another way is to use a theorem prover for first order logic and to transform the reconstruction into a first order theory by means of axiomatized set theory. Without going into the details of the theory of computational complexity, it can be said that both ways are likely to exceed the computational capacities of theorem provers currently available.[9] If mathematical expressions in the form of functions whose range is the set of real or rational numbers are needed to express scientific assertions, then theorem proving is certainly not an option. If the theory is more of the qualitative type, i.e. not requiring such mathematical functions, then an estimation of the computational complexity of the calculations required for automated reasoning with a completely formalized theory is worthy of consideration.

In light of the limitations inherent in the logic programming approach to automated reasoning, we need to seek alternative ways to represent and to process the information given by the reconstruction of a particular scientific theory. A relational database certainly does not have sufficiently expressive resources for that purpose. But using semantic markup languages seems a promising alternative to logic programming. In particular, RDFS qualifies as an appropriate means for several reasons. First, it allows for multiple inheritance. This is necessary since there are several structuralist reconstructions in which there is a theory-element that is a specialization of more than one general theory-element. Second, RDFS allows specifying relations between classes that are located in different hierarchies. This is needed to account for links between theory-elements that are components of different theory-nets. Third, RDFS provides not only a language to codify an ontology but also provides expressive resources to store data, the semantics of which are given by the ontology.[10] Hence, the data that come in the form of intended applications can be represented as entities that are related to the class of the theory-element.

It goes without saying that if a representation of a theory-element by means of RDFS has been accomplished, the reasoning capacities of an RDFS reasoner are insufficient to evaluate the experimental data of that theory-element. The use of RDFS is primarily intended to exploit the frame-like character of certain meta-theoretical concepts in structuralism. It is, furthermore, motivated by the quest for unified standards in representing experimental data and the theoretical context in which such data are to be evaluated. The evaluation itself would be the task of a procedural component of the software system, which accesses the RDFS document representing the semantics of experimental data and the theoretical context.

Another way to exploit the frame-like architecture of structuralism would be to use object-oriented programming, as supported by programming languages such as C++ and Java. Procedures evaluating scientific data would then be more closely related to *classes* that provide the data structures for representing models, potential models and partial models of the theory-elements. There would then be no separation between procedural components, which evaluate data, and components encoding the semantics of data, as well as the definitions on which the evaluation of data is based. The main advantage of using RDFS is that such a representation can be maintained more easily than object oriented code and can be made accessible through the World Wide Web in an unambiguous way.

In the scenario under consideration it has been assumed that a team of developers designs an application that evaluates data in a given theoretical context. Hence, the user does not need to know anything about the structuralist framework. As a consequence of this, the framework would be hidden behind the user interface. A potential extension of this scenario would consist in enabling the user to use the framework for the input of new data and new hypotheses. The structuralist framework or an extension of it could be used as a *modelling language* for scientific theories not only by developers but also by scientists using software applications.

Domains of Interest

So far, our discussion has remained at a rather abstract level, that is, we did not say anything about the domains that may be represented by means of the structuralist framework. In which domains may we beneficially use a semantic specification of the scientific concepts, hypotheses and the empirical systems as the subject of scientific hypotheses? To at least tentatively answer this question, let us briefly reflect on the motivation for ontological research in knowledge engineering. There are at least two major objectives that are considered to motivate that kind of research:[11]

- to facilitate the integration of semantically heterogenous data
- to build reusable components of software systems.

The integration of data can be challenging for several reasons. One common source of difficulty is the existence of different formats to represent data. Another type of difficulty arises from the task of processing sets of semantically distinct data that need to be related for the purpose of evaluation. It is the latter type of difficulty that may be addressed by using a standardization of scientific data, as developed in the structuralist framework.

Two things are widely acknowledged in the literature. First, a formal and standardized representation of biological theories would be highly beneficial to the automated evaluation and representation of data in the biological domain. Second, little has been done in formalizing biological theories.[12] In this situation, it seems worth investigating the potentialities of a semantically rich standardization, such as the one provided by the structuralist framework, to advance the representation of biological data. In particular, it is genetic information in the context of investigations into the particular function of genes and proteins that has become an issue of utmost importance in the biomedical domain. The use of microarrays has led to a vast amount of data which need to be evaluated and to be made accessible. There is work in progress by the author to present the central theory-elements of functional genomics as sequence alignment and the theory of homologous sequences, by formal means of the structuralist framework.[13]

12 ONTOLOGICAL COMPLEXITY AND HUMAN CULTURE

David J. Saab and Frederico Fonseca

Introduction

The explosion of the infosphere has led to a proliferation of metadata and formal ontology artefacts for information systems. Information scientists are creating ontologies and metadata in order to facilitate the sharing of meaningful information rather than similarly structured information. Formal ontologies are a complex form of metadata that specify the underlying concepts and their relationships that comprise the information *of* and *for* an information system.[1] The most common understanding of ontology in computer and information sciences is Gruber's specification of a conceptualization.[2] However, formal ontologies are problematic in that they simultaneously crystallize and decontextualize information, which in order to be meaningful must be adaptive in context. In trying to construct a correct taxonomical system, formal ontologies are focused on syntactic precision rather than meaningful exchange of information. Smith describes accurately the motivation and practice of ontology creation:

> It becomes a theory of the ontological content of certain representations ... The elicited principles may or may not be true, but this, to the practitioner ... is of no concern, since the significance of these principles lies elsewhere – for instance in yielding a correct account of the taxonomical system used by speakers of a given language or by scientists working in a given discipline.[3]

It is not fair to claim that syntax is irrelevant, but the meaning we make of information is dependent upon more than its syntactic structure. The semantic content of information is dependent upon the context in which it exists and the experience through which it emerges.[4] For true semantic interoperability to occur among diverse information systems, within or across domains, information must remain contextualized.

Heidegger's phenomenological examination of ontology, which includes the notion of being-in-the-world, is one in which each of us is immersed in and never separated from an experiential context.[5] This context is the ever-present background and historical experience that shapes our semantic and ontological commitments to the world and which helps us to make meaning of what we perceive to exist. Moreover, we are always being-in-becoming, experiencing the world as emergent-dynamic, contextualized and with a personal historical perspective. It is this ontological notion of being-in-becoming that allows us to introduce the notion of culture to the study of ontology in information science.

Even though Heidegger shifted the grounding of ontology in philosophy from the categorical to the phenomenological, information scientists still adhere to the notion of classification and categorization as the essence of ontology. While this may be necessary to deal with the limitations of computational systems that function primarily as symbol processors, it also constrains our ability to address the conceptual and semantic dimensions of ontology. Integral to understanding ontology, to understanding being, is the notion of background and culture – what Heidegger refers to as being-in-the-world. What exists does not exist independently of the Being that is experiencing it, nor does it exist independently of the contextual background in which it is being experienced. The clear line between subject and object, or between object and object, that exists in an Aristotelian categorical notion of ontology becomes irreversibly blurred in a Heideggerian phenomenological notion of ontology. If we are to understand being, and hence ontology as the theory of being, we must not separate ourselves from the world which is integral to our experience. If we are to understand information we must not objectify it as an entity that exists independently of ourselves. We must strive to retain the context that provides the semantic content necessary for sharing information and knowledge.

The notion of culture as described by cultural anthropologists is Heidegger's being-in-becoming – an emerging experience of the individual in a situational context. Cultural anthropologists describe culture as a phenomenon that emerges through the interplay of intrapersonal cognitive structures and extrapersonal structures in the world.[6] Culture is a phenomenon integral to our experience and one that shapes our ontological commitments to the world around us. Our experience is always a cultural experience. Our individual cognitive experience of the world is dependent in large part upon our cultural experience. What we presume to exist and the meaning that we make of the world is dependent upon our cultural schemas and experiences. Culture helps to focus our attention on and make meaning of relevant extrapersonal structures and their qualities and dimensions that comprise the context and background of the world.

Its role in the creation of meaning makes culture integral to the study of semantics and, consequently, the study of ontologies and information technologies. The meaning we make of entities and phenomena in the world is always shaped by our cultural experience. And if we understand culture as the emergent interplay of intrapersonal cognitive structures and extrapersonal structures of the world, then the notion of cognitive and cultural schemas becomes essential to understanding ontologies and the ways in which we might achieve authentic semantic interoperability among diverse information systems.

In this essay we introduce culture as an essential concept for information science. We explore the nature of ontologies and reconceptualize them as cultural schemas. We draw upon Heidegger's examination of ontology to ground ontology in a phenomenological perspective, enabling it to remain flexible and adaptable and to accommodate context.

Ontology as Category

Ontology is a *philosophia prima* concerned with the theory of being, i.e. what exists. In his *Metaphysics*, Aristotle describes ontology as regarding 'all the species of being qua being and the attributes which belong to it qua being'.[7] A ' true' ontology would be one – and there would be only one – in which all things of existence and their relationships with one another were described in a single coherent and comprehensive treatise.[8] Aristotle determined this to mean that everything could be described through a system of hierarchical categories. The historical path from philosophical ontology to computational ontologies is one that adheres primarily to the notion of ontology as a categorization and classification system. The obvious implication for ontology as categorization is that there is a single objective world that exists and that it can be described as entirely separate from the person observing it.

Interestingly, while logicians and researchers, especially from Western cultural traditions, have adopted a rationalistic world view[9] such that there exists a single objective world and that we can separate ourselves as subjects observing it, they have also recognized that different domains often have a different understanding of the same concept. This idea of differing ontological conceptualizations was described by Quine, who set forth the task of the ontologist as discerning what types of entities scientists are committed to in their theories, which are discipline-specific. This specificity means that relations among objects belonging to different domains are not necessarily compatible, resulting in multiple ontologies among the domains.[10] Ontology in the traditional philosophical sense is then replaced by domain-specific conceptualizations. This shift from ontology (singular, encompassing everything) to ontologies (plural, restricted to a particular domain), from an external to internal metaphysics, identified by Quine for the natural sciences, has found its way into the social science disciplines. Psychologists and anthropologists have attempted to elicit the ontological commitments of individuals and cultures in much the same way as philosophers of the natural sciences.[11] However, the idea that there is a single objective world separate from the persons observing it, and that humans can discern it through the study of categories, still permeates the ontological commitments of natural and social science disciplines. The multiplicity of ontologies recognizes the fact that there are varied human understandings of this objective world, not that those varied understandings constitute a multiplicity of non-objective worlds.

Though dominant, 'specification of a conceptualization' is not the only definition of ontology that exists among contemporary researchers. People interpret ontology to be philosophical, semantic, conceptual, formal, informal, representational, logical-theoretic, property-driven, purpose-driven, multi-levelled and/or see them as a vocabulary or specification. Guarino identified seven distinct ways in which people interpret the term 'ontology': (1) as a philosophical discipline; (2) an informal conceptual system; (3) a formal semantic account; (4) a specification of a ' conceptualization'; (5) a representation of a conceptual system via a logical theory, (5.1) characterized by specific formal properties, (5.2) characterized only by its specific purpose; (6) the vocabulary used by a logical theory; and (7) a meta-level specification of a logical theory. Interpretations 4–7 are the dominant interpretations for information and computer sciences. These interpretations have impact on the development, construction and use of ontologies in the wild (*pace* Hutchins).

Difficulties in practice highlight conceptual problems regarding ontologies such that, for many, ontology has come to mean one of two things: a representation vocabulary, or a body of knowledge describing some domain.[12] Ontology is the conceptualizations underlying the representational vocabulary, not the vocabulary itself; it is the conceptualizations of relationships that constitute a body of knowledge, not the description itself. Translating from one language to another, according to this view, does not alter the ontology conceptually – a transistor is a transistor is a transistor no matter whether the vocabulary representing it is in English or Farsi or Cantonese. As a body of knowledge describing a domain, ontology attempts to specify the relationships of the concepts – a transistor is a component of an operational amplifier, or that an operational amplifier is a type of electronic device. Casting ontology as categorical constrains its applicability to new information and new contexts, making the refinement or merging of these complex artefacts nearly impossible as any changes tend to break them.

One difficulty with ontology-as-category is the lack of consensus as to what those categories should be or how they should be organized. Even with as fundamental a concept as *class*, ontologists lack consensus.[13]

The most basic concepts in a domain should correspond to classes that are the roots of various taxonomic trees.[14] Concepts are terminological descriptions of classes of individuals. Concepts represent classes of objects.[15] Just as in the object-oriented paradigm, there are two fundamental types of concepts in KM: *instances* (individuals) and *classes* (types of individuals).[16] Classes represent concepts, which are taken in a broad sense.[17] A class is a set of entities. Each of the entities in a class is said to be an instance of the class. An entity can be an instance of multiple classes, which are called its types. A class can be an instance of a class.[18] The class *rdfs:Class* defines the class of all classes.[19] A class has an intensional meaning (the underlying concept) which is related but not equal to its class extension.[20] 'Instances are used to represent elements or individuals in an ontology.'[21] Individuals represent instances of classes or concepts.[22] Individuals are assertional, and are considered instances of concepts.[23]

Kuśnierczyk observes,

> The issue is not merely one of incoherent nomenclature: it is not clear whether a class of all classes, and those classes themselves, are elements of the represented domain, elements of a formal representation of the domain, or, perhaps, elements of a representation of a mental imagination of the domain.[24]

The way in which ontological engineers use *class* and *concept* interchangeably reveals the lack of clarity among those whose work is to produce precise descriptions and definitions for systems interoperability. With the lack of consensus as to what constitutes a *class* or a *concept*, which are supposedly fundamental to ontologies, it is not surprising that ontology modification and integration is a problematic endeavour.[25]

Different ontologists will categorize differently, use different descriptors to specify their conceptualizations of the things of the world. Yet despite these difficulties ontology as the study of categories persists. All categorization schemes impose a rigid or semi-rigid structure onto the entities or phenomena being described. In essence, they attempt to get back to the source of Aristotle's ontological pursuit and create categories that enable the classification and categorization of anything that exists in the world. The underlying assumption to all of these hierarchies is that they objectively describe an objective world. The plethora of structures and vocabulary within and across domains indicates, however, that the ontologies are non-objective descriptions of non-objective worlds.

While information systems are generally good at connecting incompatible systems by using or translating protocols and formats, they often fail when it comes to interpreting the meaning of specific information.[26] If the semantic content of some information does not comply with the formal ontological structure of the information system, it is not usable or interpretable by the system. The problem is one of meaning, and it is compounded by the fact that any or all of the meanings of a particular entity or phenomenon may be used by different people at different times. Smith calls this the Tower of Babel Problem: different groups, including data- and knowledge-base system designers, have their own terms and concepts for understanding and building representative frameworks.[27] Identical labels may have entirely different meanings; or the same meaning may have different labels. Information systems using different ontological classifications aren't able to communicate easily without additional layers of metadata that allow them to map one ontology to another.

Ontology as Hermeneutic

Even though the notion of ontology has continued to evolve in philosophy, information scientists still adhere to the notion of classification and categorization as the essence of ontology. Ontology from an existentialist and phenomenological perspective is not the study of categories of being but rather the study of experience of being:

> Basically, all ontology, no matter how rich and firmly compacted a system of categories it has at its disposal, remains blind and perverted from its ownmost aim, if it has not first adequately clarified the meaning of Being, and conceived this clarification as its fundamental task.[28]

The 'problem' of ontology arises from the tension between what Heidegger describes as the ontological and the ontic – between the conceptualizations we develop through experience and the instantiations of those conceptualizations that comprise entities and phenomena in the world. What is the relationship between the rich ontological understanding and conceptualizations we have based upon our experiential being and the seemingly objective ontic instantiations of what we encounter in our experience? The question lies at the heart of metaphysics and our philosophies of science, and is perhaps most critical for information science. We need to see the world as objective and distinct from ourselves, to believe that we are not solipsistic 'brains in vats' merely imagining that a world exists rather than one actually existing. We reinforce this need by objectifying nature, by imposing order on all data, to make all types of entities subject to processing.[29] We engage in this type of activity as scientists because our traditional ontological stance derives from the question, 'What *is* it?' Heidegger asks a more difficult and vexing question: what does it mean *to be*? And at the centre of this inquiry dwells man and his *being-in-the-world*.

We humans are unable to escape our experiential contexts. The essential nature of man's being is to understand, to make meaning of our experiences, of our world. Indeed, we cannot help but make meaning of our experiences, whether we are able to articulate them through language or not. We make this meaning individually and collectively. We share our experiences of being-in-the-world with others, and as such, Heidegger says, we are always *being-in-the-world-with-others*. When we share with others we do so from a personal historical perspective, from the sum total of our experiences and the understanding we have created about them. Our experiences are always contextualized and the meaning we share with others is done from this contextualized perspective. We engage one another in hermeneutic discourse where our horizons of understanding are fused with others and thereby expanded. We don't require extreme specification of one another's conceptualizations. In conveying meaning to others we make use of metaphor, imprecise language and non-verbal cues from which we can create our own, newly contextualized understanding. Our experiences are such that we are constantly negotiating the contexts in which we are immersed and sharing those experiences in imprecise ways – we are always contextualized and underspecified, as it were. Yet, we are able to understand each other and share meaningful information about those experiences.

The reasons for creating ontologies as a system of categories rest with the limitations of machine information systems. In order to communicate with one another, information systems need very specific rules and structures to share data, unlike humans who are immersed in cultural contexts. Representation is an essential feature of computation currently conceived. What is represented and how it is represented – what is considered relevant and how it is made explicit – are choices we make as information scientists. We impose order on the information to be represented, but we do so from a reflective position, separating ourselves from the world and our immersive experience. We try to abstract that experience in constructing ontology-as-category and cast it into a particular frame.

> A representation casts a frame on the world, but this frame is a strength as well as a limitation. Stepping out of the frame is like jumping out of a hoolahoop while holding it ... We can schematically classify efforts to understand the origins of representations into two approaches: induction and selection. I propose a third alternative, which relies on interaction, construction, and communication.[30]

Within computational systems, a frame provides the structure within which information can be exchanged. Structure is necessary for meaningful exchange, but it is insufficient. We might speak the same language and use the same syntactic rules, but that doesn't guarantee meaningful communication. We might make different ontological commitments that employ different assumptions, which we uncover through hermeneutic discourse – interaction, construction and communication. Our discourse occurs in context and to which we bring an 'existential *fore-structure*' as part of the hermeneutic 'circle' of understanding:

> [u]nderstanding always pertains to the whole of Being-in-the-world. In every understanding of the world, existence is understood with it, and *vice versa*. All interpretation, moreover, operates in the fore-structure ... Any interpretation which is to contribute understanding, must already have understood what is to be interpreted.[31]

Ontology-as-hermeneutic is a reflection of our experience of being. We don't enter the hermeneutic circle as blank slates, rather we carry as part of us the 'existential fore-structure' necessary to understand, and through which we co-create meaning as we interact with one another. The fore-structure is developed from and integrated into our personal, historical experience. And because we are being-in-the-world-with-others, the fore-structure becomes part of our shared background, our shared understanding of the contexts in which we interact. Heidegger's fore-structure is what we describe as a network of cognitive schemas, some of which we share as cultural schemas. They are experiential and emergent rather than analytically reflective. They are continually evolving, adapting and changing, though not chaotic. Rather they are relatively stable, but with tolerance for variability and ambiguity. In taking a perspective of ontology-as-hermeneutic, our ontologies *are* our cultural schemas.

Cognition and Cultural Meaning

In cognitive science, connectionist theory posits the human conceptual system as a network composed of a large number of units joined together in a pattern of connections.[32] Cognitive anthropologists and educational psychologists refer to these patterns of connections as schemas.[33] Schemas are strongly connected networks of cognitive elements, having a bias in activation through repeated exposure to the same or similar stimulus, but they are not rigid and inflexible.[34] D'Andrade explains in more detail that schemas are 'flexible configurations, mirroring the regularities of experience, providing automatic completion of missing components, automatically generalizing from the past, but also continually in modification, continually adapting to reflect the current state of affairs'. Describing them as 'flexible, mirrored configurations' implies that schemas are structural entities within cognition that are comprised of several elements. Schemas are not the individual elements, but rather strongly connected clusters of elements of experience within cognition. Elements of experience are clustered in cognition, in our neural networks, because they are clustered in our lived experiences. Schemas are cognitive entities that help us process information. Clustering cognitive elements makes them more efficient by reducing the cognitive load associated with processing experience.

Schemas also have other qualities. Some schemas are durable. Repeated exposure to patterns of behaviour strengthens the networks of connections among the cognitive elements. Some schemas show historical durability. They are passed along from one generation to the next. Some schemas show applicability across contexts. We draw upon them to help us make sense of new and unfamiliar experiences. Some schemas exhibit motivational force. Such motivation is imparted through learning, explicitly and implicitly, strengthening the emotional connections among the cognitive elements. Schemas have a bias in activation through repeated exposure to the same or similar stimulus, but they are not rigid and inflexible. They are adaptable, sometimes resulting in the strengthening of existing schemas, sometimes in their weakening in the face of new experience.

Schemas are the cognitive element in the 'structural coupling' of our experience as described by Winograd and Flores.[35] Schemas are powerful processors of experience, help with pattern completion, and promote cognitive efficiency. They serve to both inform and constrain our understanding of experience. Because of their functionality in pattern completion, schemas function – in some sense – as flexible filters of experience, enabling us to attend to its salient features while filtering out the non-salient. People recall schematically embedded information more quickly and more accurately.[36] In fact, schemas hold such sway in our cognition that people may falsely recall schematically embedded events that did not occur. They are more likely to recognize information embedded in existing schemas because of repeated activation of the patterned cognitive elements.

Repeated activation evokes expectations within cognition, and the easy recognition or dismissal of contradictory or challenging information that do not conform to those expectations formed as part of the existing schemas. Information that is orthogonal to existing schematic structures, that doesn't acquire salience through the repeated activation of schemas and the creation of associated expectations, is much less likely to be noticed or recalled. Unless, of course, orthogonality becomes the focus of the experience such as when we are working to expand our horizons of understanding through discourse. In this situation we are attuned to the divergence between our shared schemas as we try to close the distance between our conceptualizations and those of others. Or, as Heidegger might say, we are always attending to the breaking down of experience.

Schemas, as complex cognitive associations, are intrapersonal structures. The objects or events that are manifest outside individual cognition, the entities in the external world, are extrapersonal structures. It would be inaccurate to say that schemas are separable from culture, for that would imply that culture consists solely of the external world structures outside the individual. Culture consists of the interplay between the intrapersonal cognitive structures – Heidegger's existential fore-structures – and extrapersonal structures such as systems of signs, infrastructure, environment, social interaction, and so on – Heidegger's *ready-to-hand* background. The intrapersonal and the extrapersonal are different and distinct, but closely interconnected. They are not isolated from one another, rather separated by a permeable boundary – one that blurs the distinction between subject and object. Culture encompasses both intrapersonal and extrapersonal structures and emerges from the interplay between them. It is through this interplay that we can see that some of the intrapersonal cognitive structures called schemas are shared. The notion of schemas marks a shift away from the focus on deliberative and explicit cognitive processes, which mirror the ways we deal with language in cognition (and formal ontology artefacts), to thinking and cognition as automatic and implicit. It is the shift away from symbolic processing models of cognition toward a connectionist model of cognition, a shift from ontology-as-category to ontology-as-hermeneutic.

Shared Schemas as Cultural Schemas

The sharing of schemas does not require people to have the same experiences at the exact same time and place, rather that they experience the same general patterns. As beings in the world, we organize our experiences in ways that ensure ease of interaction, coordination of activities, and collaboration. Because we organize our experiences in particular ways, people in the same social environment will indeed experience many of the same typical patterns. In experiencing the same general patterns, people will come to share the same common understandings and exhibit similar emotional and motivational responses and behaviours. However, because we are also individuals, there can be differences in the feelings and motivations evoked by the schemas we hold: 'The learner's emotions and consequent motivations can affect how strongly the features of those events become associated in memory'.[37] Individuals will engage with the external world structures and experience the same general patterns. Similar stimuli and experiences will activate similar schemas. It is in this sense that we consider them shared schemas. It is their quality of sharedness that makes them a dimension of the cultural.

We share the intrapersonal dimensions of culture when we interact with others. In sharing these intrapersonal dimensions, schemas are activated. Activation evokes meanings, interpretations, thoughts and feelings. The cultural meaning of a thing, which is distinct from the personal cognitive meaning, is the typical interpretation evoked through life experience, with the acknowledgement that a different interpretation could be evoked in people with different characteristic life experiences. In some cases our experience is intracultural, where we share a similar cultural frame. In other cases our experience is intercultural, where we are sharing different cultural frames. The meanings evoked by one person in relation to a particular extrapersonal structure may not be the same as those evoked in another. In fact, the meanings evoked may not be the same within the same person at different times, for they may experience schema-altering encounters in the interim. The ways in which we share these intrapersonal dimensions of culture makes each person a junction point for an infinite number of partially overlapping cultures.

The Importance of Identity

The notion of identity and multiplicity of perspectives is important in our understanding of how cultural schemas manifest. Individuals can manage multiple identities in the same or multiple contexts. We can shift our perspective effortlessly between national, familial, peer and other identities to make sense of particular phenomena (i.e. frame it in relation to ourselves).[38] The same context, for example, that would be considered 'exciting' to 'the hunter' might also be ' dangerous' to ' the parent'. Fauconnier and Turner claim that 'frames structure our conceptual and social life and, in their most generic and schematic forms, create a basis for grammatical construction'. Words are themselves viewed as constructions, and lexical meaning is an intricate web of connected frames. They also claim that although cognitive framing is reflected and guided by language, it is not inherently linguistic – people manipulate many frames even though they have no words and constructions for them. It is the individual's salient, contextualized identity in relation to the phenomena that allows for sensemaking of the phenomena. When making meaning of a particular phenomenon, individuals will rely upon the cognitive and cultural schemas that are integral parts of their salient, contextualized identities.

Knowledge embodied in conceptual systems and reflected in language is in fact deeply ensconced in culture.[39] Our interpretation of reality is dependent upon our cognitive structures and our cultural and contextual backgrounds; all associations of perceptual input to cognitive concepts depend on our pre-understanding of the context.[40] Our cognitive schemas shape the associations we make of our perceptual input.

Culture and Ontological Commitments

The construction of ontologies by information scientists is an attempt to overcome the Tower of Babel problem by providing a common dictionary of terms and definitions within a taxonomical (i.e. relationship) framework for knowledge representation that can be shared by different information-systems communities.[41] However, theories of being, of what exists, are not defined by a common vocabulary, rather they are dependent upon particular perspectives and ways of understanding the world in which we are immersed. What exists is dependent upon our cultural schemas. Recognizing the diversity of understanding and ways of creating and communicating meaning about our world is not an anti-realist position – in fact, we consider this recognition to be a solidly realist position. Ontology-as-hermeneutic does not dispute that there exists a universe in which we are all embedded and able to experience or even that we can come to similar understanding from radically different cultural traditions. Rather our perceptions, understandings and meanings are co-created through our cultural interactions. Ontology-as-hermeneutic simply does not *a priori* privilege one ontology – the one that just happens to correspond to our Western cultural traditions – over another.

We can illustrate this point with a simple, yet high-contrast, question: is that rock mound in the West Australian desert a granite composite with specific Cartesian dimensions or is it Krantjirinja, my Kangaroo Ancestor? The answer to that question is dependent upon the cultural schemas of the person being asked. For the geographer, the mound is a rock formation, composed of slate and situated on a Cartesian grid, which he can map and represent in a GIS. For the Krantji, an Aboriginal group on whose land the mound is situated, it is Krantjirinja, their Kangaroo Ancestor, who has existed since and continues to exist as part of the Dreamtime, and who sits along a path commonly known to outsiders as a songline. And they see not only Krantjirinja but also his influence and power on the surrounding landscape. The Krantji clan does not see a rock formation when they look at the mound – it doesn't exist – just as the geographer doesn't see Krantjirinja. Not only are identified geographic entities constructed culturally, but the relationships between entities in geographic space also are associated in different ways by different cultures.[42] In making meaning of this mound, each uses the cultural schemas they have developed through their cultural immersion. They literally see completely different entities situated in completely different backgrounds.

What happens if a member of the Krantji clan becomes a geographer? To become a geographer one must have a particular type of training or education. One does this through a process of acculturation, where one is exposed to and assimilates the ideas, concepts and understandings of other geographers. The Krantji clan member doesn't come to the acculturation process devoid of experience or world view, rather with a set of cultural schemas she uses to understand the geographic landscape. In learning the new cultural schemas of geographers, she will acquire new cultural schemas and possibly blend them with her Krantji cultural schemas.[43] Or, she could compartmentalize the two sets of cultural schemas, which would be evoked in different and separate experiential contexts.[44] Once she has acquired the cultural schemas of geographers, she may be in a unique position to translate between the ontologies of the two cultures. Perhaps she discovers that what connects the two ontologies is not the outward appearances of a rock mound and a Kangaroo Ancestor, but its location as the primary source of water and the determinant of the hydrologic cycle of the area – important to the cultural schemas of the physical geographer – as well as those of the Krantji clan who think of it in terms of the power and influence of their ancestor, which is intimately linked to ecological conservation and survival. The rock mound and the ancestor may simply be the entry points into the complex ontological associations that we as humans engage in a hermeneutic process to uncover. Employing ontology-as-hermeneutic allows us to create bridges between sets of different cultural schemas – different ontologies – to create a new shared cultural schema that the bicultural Krantji geographer would be in a unique position to facilitate.

Cultural schemas are, in essence, our ontologies. They shape our ontological commitments to what exists in the world as well as the ways in which we approach and engage with the world. And while they help structure our understanding of the world in which we are embedded, they are associative and flexible. They help to focus our attention on particular details of our experiences and give them salience. They allow us to make meaning of the contextualized, cultural experience in which we are always immersed. Formal ontologies constructed as taxonomic structures and categories of an objective world, however complex and inclusive of relationship axioms, will not work across cultural boundaries because they rest on different ontological conceptualizations and commitments. They crystallize a single perspective into the ontology artefact as representative of what exists. They short-circuit the dialogue that humans engage in as part of their semantic negotiations about their ontological commitments. Moreover, they work only in limited degree across individual or intracultural boundaries. Humans think and communicate in very flexible and schematic ways, and ontologies should reflect this flexibility and the adaptive nature of human cognition in order to achieve semantic interoperability. In order to do so, we must forego the comfort of a rationalist world view that presumes an objective external world as well as its logical opposite, solipsism. We need to reach beyond the lexical and syntactic in constructing our machine ontologies that rely on symbol processing and extend their grounding to the phenomenological and hermeneutic – embed within them the ability to negotiate meaning through a hermeneutic process. Casting culture as an emergent phenomenon, and cultural schemas as the complex networks of conceptualizations that comprise our ontologies, allows us to ground ontologies on a phenomenological footing.

13 KNOWLEDGE AND ACTION BETWEEN ABSTRACTION AND CONCRETION

Uwe V. Riss

Introduction

The management of knowledge is considered to be one of the most important factors in economic growth today.[1] However, the question of how to deal with knowledge in the most efficient way is still far from answered. We observe two fundamentally different approaches to the question of how we should deal with knowledge. One view sees knowledge as a kind of static object that can be gathered, compiled and distributed; the other view regards knowledge as a dynamic process.[2] This disaccord finds a parallel in an objective–subjective distinction where the first position sees knowledge as independent of personal opinion whereas the second position regards it as interpretative.[3] These discussions are not merely academic but crucially influence the way that knowledge management (KM) is realized, i.e. whether the focus is placed on knowledge artefacts such as documents or on subjective acts.

The particular interest of the current essay concerns the question of how KM can be supported by information technology (IT) and which are the fundamental structures that must be regarded. Traditionally, IT-based approaches favour an object-oriented view of knowledge since knowledge artefacts are the objects that can be best processed by IT systems. This even leads to the view that knowledge artefacts represent the only form of knowledge.[4] On the philosophical side this perspective is fostered by analytical investigations that emphasize the primacy of propositional knowledge that is closely related to knowledge artefacts.[5]

However, several recent studies support the view that knowledge possesses a dynamic nature that cannot be easily transformed into knowledge artefacts.[6] This raises the question which of these views (static or dynamic nature of knowledge) is true or how they are related. In relation to the dynamic aspects of knowledge especially the connection of knowledge and action has been brought into the discussion as one of the central topics to be investigated.[7]

The parallel existence of static and dynamic views of knowledge has raised the question of whether a perspective that integrates both views might be a reasonable approach. Thus, Nonaka and Toyama have brought up the idea of a synthesis of both views applying dialectics.[8] Unfortunately their investigation has only touched the surface of this thoroughly philosophical question which requires some insight in the foundation of dialectical thinking.

In the following investigation we want to provide this deeper understanding of the relation between knowledge and action and discuss in which respect the explanation of knowledge and KM approaches can benefit from dialectical thinking that includes action-related aspects. To this end we start the investigation with the concepts of abstraction, which we introduce in the second section (pp. 146–9), and concretion, discussed in the third section (pp. 149–55). While the former refers to the static view, the latter represents the dynamic aspects. It is argued that both are dialectically related and that every abstract description of the world is always limited. These limitations are due to the fact that abstraction falls short of coping with concrete processes. In the fourth section (pp. 155–7), the concept of capacity and its relation to abstraction and concretion is discussed. In the fifth section (pp. 157–62), we finally come to the concepts of knowledge and action. It is argued that action brings together different acts of abstraction and concretion and that the capacity related to action is a rational capacity that accords with the knowledge to control the respective action.[9] It is argued that knowledge and action are connected in the meta-process of learning as a sublation of knowledge and action. Moreover, it is shown how the analytical standard analysis of knowledge is related the concept of knowledge as rational capacity. In the sixth section (pp. 162–6), we finally give some practical consequences of the presented view with respect to KM. This particularly concerns the inclusion of action representation in KM systems. In the seventh section (pp. 166–8), we give a short discussion of the results and related issues.

Abstraction and its Limitations

Although 'abstract' and 'concrete' are frequently used terms their meaning is not always sufficiently clear. This especially concerns the question of what is meant by 'concrete'. Therefore we will first give an explication of both starting with the term 'abstract' which appears to be less fuzzy. In both cases we follow the investigations of Ruben.[10] He states that an object is regarded as abstract if it is exclusively considered as a representative of an (equivalence) class to which this object belongs. According to Ruben,

> abstraction denotes the transition (process) from a practical treatment or theoretical consideration of different but equivalent objects (or ensembles of objects) of a referential domain to a consideration of these objects (or ensembles of objects) as

representatives of an (equivalence) class with respect to which these objects (or ensembles of objects) are equivalent.

By objects we do not only mean material objects but everything that can be subject to such classification including events (e.g. volcanic eruption, election), activities (e.g. walking, thinking), state changes (e.g. evaporation, dying), etc.

Abstraction imposes a distinction that separates objects that belong to a specific class from those that do not. It is a common assumption that this separation is based on specific object features as in the example of the term 'bachelor' defined as a 'single' 'man'. This definition is based on a restricted set of essential features although a concrete bachelor might also possess other features, e.g. he might be blond or bearded. The assumption is that there is always a unique set of features that characterizes an object as representative of a specific class in an unequivocal way. This view is expressed in set theory where a set A is usually described in the form $A = \{ x \mid PA(x) \}$, i.e. as the class of all objects x that possess a property $PA(x)$. We can regard this as the mainstream view with respect to abstraction.

However, Wittgenstein has shown in the discussion of family resemblance that such a reduction is the exception rather than the rule.[11] In general it is simply not possible to determine a unique set of essential features. This suggests that the required identification cannot be derived by analytic reasoning. It is, rather, the result of a concrete (physiological and mental) process which essentially includes the interaction of the identifying person with the external (concrete) object in a concrete situation. If we imagine a situation in which two persons stand in front of a chair, it might happen that the person P1 asks a second person P2 what this object is. P2 may immediately answer that it is obviously a chair. However, if P1 now asks how P2 knows this, generally P1 will be astonished and is likely to point out that this is obvious. This indicates that in general abstraction does not take place as an analytic (logical) mental process in which a person explicitly checks all features of the objects but in an implicit and subsidiary way. In accordance with Ruben and others we conclude that identification must be regarded as an act that is based on a fundamental practical and theoretical capacity instead of a given property determined by unique features.[12] The particular capacity, on which such an abstraction is based, is mostly acquired in a process of learning. For example, we learn how to identify chairs and tables and often we do so mainly without explicit reference to definitions. We will return to this aspect later.

Abstracts can appear in various forms. First of all they appear in an atomic form as concept, e.g. the 'chair' as a perceived object. However, human beings can also recognize abstract connections such as a 'brown chair' or a 'brown chair standing in the living room', or entire propositions such as 'The brown chair is standing in the living room'. Therefore we consider propositions as abstract connections. The latter point might pose the question of how identification is understood in this case. Let us take the proposition 'The apple fell to the ground'. Obviously this proposition is an abstract connection and identification must be understood in the sense that an observer correctly states this proposition in a situation to which it is applicable, e.g. while standing in front of an apple tree. In other words, the identification is an actualization of the capacity to identify situations to which this proposition is applicable. A common feature of all abstraction is that they are assigned to truth values with respect to the situation in which they are used. It is true or false that the object on which I sit is a chair, in the same way as the proposition that describes this is true or false.

In abstract connections we find several abstractions at the same time. From the logical point of view such abstract connections are often related to set theoretic operations. For example, if $C = \{ x \mid PC(x) \}$ denotes the class of all 'cherries' and $R = \{ x \mid PR(x) \}$ the class of all 'red objects' then the class of all 'red cherries' is $CR = C \cap R$. However, here we find the first limitation of abstract thinking if we refer to the experience that the whole is more than the sum of its parts. Ruben pointed to this difference which indicates that the connection between different features is not only external as in the mathematical representations. For example, if we talk about red cherries we do not only refer to the fact that the perceived objects are cherries and that their colour is red but implicitly also to the fact that this means that these cherries are likely to be ripe and tasty. When we identify 'red cherries' we take this into account and not only the fact that the perceived objects are red and cherries. Kern asserted that it is this deficiency that mainly causes the problems in the analytic discussion of knowledge. The standard analysis sees knowledge as justified true belief.[13] This leads to problems if we consider the three features 'truth', 'belief' and 'justification' as external to each other as Gettier has shown.[14] His example of justified true belief that is not knowledge exactly points at the missing internal relation between truth, justification and belief. If identification, however, requires more than identifying features, the identification process definitely demands more attention, as we will show.

If we talk about missing internal relations between abstract features we often refer to interactions that appear in processes. Let us consider the example of the construction of a house. We can ask at which point of time the house actually comes into being. Either we have to admit that we cannot determine this point or we have to artificially fix an arbitrary point. For example, if we fix it by taking the roofing ceremony or the handover of keys as the point at which the house comes into being, we have to deal with the fact that the actual object is not significantly different at the moments closely before and after the event. The reason is that the process is mainly continuous. The bivalence of abstraction, i.e. reflected in the view that the object is either a house or not, does not correspond to the actual construction process in which the house comes into being bit by bit. The attempt to clearly distinguish between both cases always remains largely arbitrary. We find similar deficiencies of abstraction in the so-called sorites paradoxes such as the heap paradox.[15] If an object is a heap of sand and we remove one grain the remaining object is still a heap. However, if we continue with this procedure we will reach a point at which the heap will obviously cease to be a heap. Also, here we have an abstract object – the heap – that undergoes a process of continuous change, namely the reduction of grains. Although the individual steps of this process do not seem to change the character of the object that still remains a heap, we find a qualitative change of the object at the end of the process when only one or a few grains are left. Obviously we need a fundamentally different way of description to understand transitions of this kind. The examples show the same systematic defect related to abstraction that we can generally construct in all kinds of abstract settings.

In the following we want to apply dialectics to deal with such deficiencies of abstraction. In particular we see a necessity for dialectics exactly in these inadequacies of abstraction. However, this requires a better understanding of the nature of processes. We have to take into account that the observed contradictions, as in the case of the heap, are of a different nature than the logical contradictions that we know from mathematics. While analytical consideration deals with internal contradictions, i.e. within an abstract framework of objects that are considered as external and independent of each other, dialectical thinking deals with external contradictions, as in the heap paradox where the contradiction results from the abstract description of a quasi-continuous process which requires an external point of view. It is to be remarked that of course we can also introduce an abstract continuous description of the process but still this cannot resolve the principle separation between heaps and non-heaps. For example, there are also abstract descriptions of continuous processes in physics and elsewhere. However, Ruben argued – based on the example of motion – that such formal (abstract) descriptions of continuous processes do not resolve the principle problem related to the Law of Logical Contradiction.[16]

Obviously these problems of abstraction do not cause any major problems in everyday contexts. However, they essentially concern the concept of knowledge and therefore practical KM applications. In the following we will discuss how we usually deal with this issue in concrete processes.

Concretion and Dialectics

The central question that we want to address in the following is how to overcome the deficiencies of abstraction. To this end it is necessary to realize that we must consider objects in their interaction with other objects instead of considering them in isolation. This means that we have to proceed to a *concrete* consideration of the object. As Ruben explained there is a widespread misunderstanding of what a concrete consideration of an object looks like. Often the term 'concrete' is used in the sense of a more specific but still abstract description. For example, we say that we describe a cherry in a (more) concrete way if we describe it as red and tasty and juicy rather than simply as a cherry. However, a cherry is not concrete because it is perceiv*able* or manage*able* in a specific way but by the actual perception or management, i.e. the concrete interaction. Ruben expresses this in the statement that the concrete is not *described* but *done*.

In the following we want to examine how this *doing* is to be understood. We have to consider the development of an object that gets involved with a subject in concrete interaction. In the same way as the abstract is the result of the process of abstraction, the concrete is the result of the process of *concretion*.

First of all, taking the interaction of objects into account means to consider objects in singular situations, i.e. we take a historical perspective towards the objects. For example, referring to the concretion of a house we have to consider the act of its construction. Even concrete processes can be described in abstract terms. However, concrete process are essentially contingent. We rather have to take the standpoint of a historian and describe concrete processes from a historical point of view, being aware that the used abstract terms refer to singular events.

Apart from cases of mere unconscious interaction of matter, the main processes of concretion include an acting subject (agent). It is the agent who performs the abstraction that actuates concretion. Even if concretion often refers to obviously physical interaction it not restricted to physical manipulation but also includes thinking and reflection as processes taking place in time; physical work must be considered in the same way as intellectual work. In the concrete process the interaction goes beyond the restrictions of abstraction; the interaction is not restricted to the aspects included in abstraction. The abstract objects become concrete by overcoming the isolation presupposed in abstraction. The key feature of concretion is the resolution of the isolation introduced by abstraction.

Concretion can be also understood as part of an evolutionary process. On the one hand, it is contingent. For example, a reflection on some concept might bring up this or that idea by chance, in the same way as the construction of a house can be influenced by various factors that make it necessary to deviate from the original plan. Quantum and chaos theory can explain why this is the case; uncertainty is a factor of human beings' everyday life. However, concretion also brings about some general aspects. For example, an individual measurement in a physical experiment usually does not allow for general statements due to statistical effects in the measured results. The repetition of the same experiment, however, usually reveals significant trends that can be described in general terms again. Probability theory can be seen as the attempt to deal with such uncertainty and to restore certainty at least partially, e.g. by providing specific distributions. Probability theory provides static abstract descriptions. Also, in evolution we observe specific trends to more complex forms of life or reoccurring patterns in life forms, independently of the contingency of individual mutations.

We even find abstract and concrete perspectives in mathematics, e.g. in mathematical theorems. The abstract aspect of a theorem is that it is true, independently of the fact of whether someone has succeeded in finding a proof. We would never say that the theorem became true by its proof but that it has been true all the time and will remain true in future as well. From a practical point of view, the mathematical theorem has a history, i.e. its proof has been found and acknowledged at a certain point of time by some mathematician. In particular we cannot make mathematical use of the theorem (even if it is actually true) before the proof has been accomplished. Before it has been proved its influence might be restricted since it cannot be used in other proofs or only as a hypothesis. The completion of the proof is a *historical* act and gives the content of the theorem a different meaning. From the logical (abstract) point of view the truth of the theorem is the same before and after the proof. Dialectics considers mathematics from its historical perspective including the actual activity of mathematicians. This is consistent with the statement that the concrete is what is done since the concrete description refers to the individual act.

Looking at the relation of abstraction to concrete objects we have to keep in mind that abstraction – the act of identification – is often influenced by concrete goals set by the agent. Whether a concrete object, e.g. a tree that we see in front of us, appears to us as a living being or a piece of wood can strongly depend on the agent's interests. This does not foster arbitrariness but simply reflects the fact that concrete objects allow for different views. Abstraction does not imply that we can describe a concrete object in only one way; it only forces us to use it in a consistent and appropriate way. It is not surprising that Hegel wrote the following about the concrete: 'the manifold or diverse is in a state of flux; it must really be conceived of as in the process of development'.[17] Let us assume that a person possesses a little table which she rarely uses. Some day she needs a stool and realizes that the little table is quite suitable for the respective purpose. Thus she decides to use it as a stool and it works. The fact that the object is used as a stool is a historical fact. There is also no necessity to use the objects as a stool. The way in which a person uses an object is not a question of logic but of needs and practicability. Obviously there is some variability in how we use and describe objects. On the other hand it is not arbitrary, since we cannot use the table as food.

The example shows that abstraction reflects a specific attitude towards the object. If a person has to decide whether she can eat a specific object, she must be able to identify this object as food and distinguish it from inedible things in a process of abstraction. This view determines the conditions for the abstraction by a specific person. In the case that the person is looking for something to eat she will identify an apple in a different way than in a situation in which she is looking for entertainment; in the latter case she might consider the apple as a ball to juggle with instead of food. Both perspectives are valid and do not cause any logical contradiction. Different situations simply bring forward different features of an object according to the respective requirements.[18]

Concretion requires a historical view that can reveal features that are not comprehensible from a logical point of view. If we take the heap paradox as an example, we find that the heap character is conserved as long as the grains are arbitrarily (in a statistical sense) removed and added. Even if grains are systematically removed this does not lead to a logical contradiction; the result can still be interpreted as a heap. However, the systematic removal of grains reveals some characteristic *tendency* that changes the quality of the object. Such tendencies are an indicator for development and often related to the emergence of new qualities that lead to the contradiction as the sorites paradox shows. Ruben calls them *concrete contradiction* since they are not due to internal logical inconsistencies.[19]

The specific way in which dialectics deals with concrete contradictions is *sublation*. Sublation is related to the development of systems showing a specific tendency that leads to concrete contradictions related to new qualities. The proceeding in sublation is described in an Hegelian way by (1) *resolving* the concrete contradiction by development of the affected abstracts, (2) *preserving* the still valid aspects of abstract objects in the concrete process, and (3) *lifting* the abstraction to a further developed level that better fits the observed concrete process.

Sublation cannot be understood from an analytical (logical) perspective but only by reference to concrete processes. Since sublation is not an analytical act, it cannot be validated in a formal but only in a practical way. In order to understand this assertion we can refer to the sorites paradox. Let us assume there are two philosophers, one starting with a heap and another starting with a single grain (non-heap). By logical analysis they now start to describe the different language games that result from virtually removing and adding grains, respectively. Following an abstract and logically correct argumentation they will finally end up with different results for the same number of grains: whereas the first philosopher rightly states that the resulting object is a heap, the other rightly states that it is not a heap. Both philosophers have followed logical rules and have good arguments for their positions. What both have neglected is to think about the principle limitation of their argumentation. They have forgotten to think about the tendencies of their proceeding and their consequences; they have not thought it through to the end. In the heap paradox, dialectical thinking has helped us to understand that we deal with a concrete contradiction and not a logical contradiction and that such concrete contradiction cannot be resolved by logical means.

Indeed we have come to know some general features of processes. Theoretical dialectics points at some general patterns that we can observe in these processes: the *Law of Dialectical Contradiction*, the *Law of Transition* (from quantitative change to qualitative change), and the *Law of Negation*. We find the Law of Dialectical Contradiction in every process where the objects simultaneously change and remain the same, e.g. the 'heap' that keeps it identity as an object during the process but obviously undergoes continuous change. The Law of Transition describes that the systematic removal of grains leads to a qualitative change of the heap although the individual steps are quantitative, i.e. not changing the quality. Finally, the Law of Negation states that the heap is negated in the process of removing grains, i.e. the heap becomes a moment of the process and is resolved in this way, and ends up in another form as a non-heap that only consists of a few grains or no grains. The central point is the resolution of abstract objects in the process and the emergence of new abstract objects.

In the following we want to consider some examples from other areas that might help us to better understand the central nature of sublation. Moving to physics, Pietschmann has argued that the quantum mechanical concept of wave–particle duality can be understood as a sublation of the two concepts of wave and particle which has been accepted due to its better fit to the microscopic experimental results as compared to classical mechanics.[20] That the derivation is not an analytical process can be easily recognized from the fact that the axioms of quantum mechanics are not mathematically derived from classical mechanics but postulated in a speculative way and validated by experiments. The creation of new abstractions is a historical process of creativity. In this sense sublation is always connected to an act of creativity even if the development is not accidental, in the same sense in which the work of an artist is creative but not accidental, i.e. it usually follows a certain historical development of the artist or an artistic movement.

An example which illustrates that sublation occurs even in the realm of mathematics is Russell's antinomy.[21] Naïve set theory allowed the construction of classes such as $R := \{ x \mid x \notin x \}$. However, this class reveals a logical contradiction, namely $x \notin R \Rightarrow x \notin x$. This would mean that $R \notin R \Leftrightarrow R \notin R$, i.e. the logical contradiction. From a dialectical perspective it was the discussion process of naïve set theory that led to the discovery of this logical contradiction. Therefore we can also describe it as a concrete contradiction so that in this case the logical and concrete contradictions coincide. The problem was resolved by including an additional axiom that excludes exactly this type of set formation. From an abstract point of view the former axiomatic system was incorrect while the augmented system was correct. However, apart from the exclusion of exceptional cases such as the above-mentioned class R, the general characteristics of set theory did not change with the introduction of the new axiom, so that the central character of set theory was preserved and we can regard this development as a sublation initiated by the historic process of Russell's discovery of the antinomy.

To answer the question why abstraction and concretion are important for us we can say that they address different needs. Abstraction must be seen as a solution to the problem of how to find stable structures in the world in order to predict the future despite the contingencies that we find in our experience. Predictions require stability of valid representation which can only be achieved on the basis of general and law-like descriptions. The predictive character is also the reason why abstraction does not allow for logical contradictions – these would make our predictions arbitrary; in logic it is well known that any arbitrary assertion can be derived from a false proposition. Turning to the role of concretion we see that in our interaction with the world we face the problem that our predictions do not always come true. Moreover, our environment (or at least some aspects in it) is constantly changing and we have to adapt to these changes. This cannot be accomplished on the basis of eternal laws and representations. Popper already pointed at the particular role of experiments in this respect. Although they cannot prove laws of nature we need them as standard to falsify these laws.[22]

However, it is well known that Popper did not overly esteem dialectical thinking. Actually, he regarded it as a 'particular variant of the trial and error method'.[23] Although he admitted that an old theory might be a good approximation to new theories he did not see this as a development but as a complete break. This is also the reason why he emphasized the aspect of falsification and not that of development. He mainly ignored the preservative character of falsification as part of sublation. Thus, most of Popper's critique of dialectics can be put down to misinterpretation. If he criticizes that our mind, i.e. the actual work of a scientist, produces sublation and not the contradictions as such, then this is completely in accordance with the view that we have presented here. In his criticism that it is misleading that sublation 'preserves the better parts' of the contradicting statements, he misinterpreted preservation. For example, the laws of classical mechanics and quantum theory are fundamentally different; nevertheless there are various touch points, e.g. how to measure particular physical values. The main point, however, is that he did not distinguish between logical and concrete contradiction. Following the explication of dialectics by Ruben it is incorrect to say that 'dialecticians claim that [the law of contradiction] must be discarded'. However, referring to trial and error as the method of science, he cannot explain how falsification actually leads to scientific progress. The development of new theories is due to genial scientists but not to tendencies in scientific thinking if we regard science as a continuous process. In reducing the consideration to trial and error, i.e. the resolution aspect, Popper ignored the preservation aspect of sublation. If we look back into the history of science we cannot generally say that scientific revolutions have devaluated the entire scientific experience (including the scientists' skills) but mainly led to a revision of the theoretical (abstract) descriptions while the experimental (concrete) methods often remained stable. We will come to this question in the next section.

Dialectical thinking corresponds to our everyday experience that even if we fail in some of our actions the main part of our experience still remains valid and that we learn from our errors. Let us take quantum mechanics as an example. From an abstract point of view quantum mechanics falsifies classical mechanics, i.e. if we consider quantum theory as true, classical mechanics is simply false. However, even if it is false we still use it where it is appropriate, e.g. to build bridges. The general argumentation is that in certain cases classical mechanics is a good approximation for quantum mechanics. However, the crucial term in this respect is 'approximation'. It suggests that classical mechanics is 'not completely false' but this statement is not compatible with the Law of Contradiction which can only distinguish between true and false, i.e. the expression 'not completely false' does not make any sense from a logical point of view. From a dialectical point of view quantum theory as the sublation of classical mechanics preserves certain features – it even confirms the validity of classical mechanics in suitable fields – and it resolves contradictions that appeared in the application of classical mechanics in the microscopic realm. Sublation does not replace false abstracts by true abstracts but it points to the limitations of existing abstracts and further development. If we refer to one reality, then different descriptions of this reality must be consistent and intrinsically connected. This is exactly the paradigm of dialectical thinking in contrast to pluralism which states that different independent views of reality might exist and even lead to contradicting descriptions. In contrast dialectical thinking states that concretion is the final point of convergence of all views. Dialectics does not allow that classical and quantum mechanics yield different answers to the same question but requires *us* to decide where classical mechanics comes to an end and quantum mechanics has to take over.

Dialectical development resembles evolution. In the same way as evolution it does not allow for a prediction of the future but provides a schema for interpreting development. Dialectic theory points at the stable features in historical development and the insight in dialectical structures helps us to deal with problems that arise from the abstractions that we use (and must use).

Abstraction, Concretion and Capacities

In the following we will further investigate abstraction and concretion and their relation to agents. Abstraction includes identification that starts with a concrete object and ends up with the assignment of this object to an equivalence class.

Concretion starts with an abstract to be concretized and ends with a concrete object as the result of this concretion. Both processes are controlled by agents and we have to ask for its conditions.

To perform the acts of abstraction and concretion the agent relies on corresponding *capacities*, i.e. the power or ability to perform these acts. A capacity describes dispositional properties of an object, e.g. the capacity of a bell to ring, of a person to swim, of sugar to dissolve in water, etc. A capacity can be natural, as in the case of sugar that dissolves in water, or acquired, as in the case of a person swimming. Kern describes a capacity as something general or abstract since it is always related to a class of acts that actualize this capacity. In this sense capacities contrast to states that are time-dependent. The abstractions as well as concretions performed by an agent obviously actualize capacities. In the case of abstraction and concretion the capacities are rather related to a particular behaviour than a controlled act due to the subsidiary awareness of both processes.[24] For abstraction this becomes manifest in the fact that abstraction takes place spontaneously and intuitively. To give reasons for a specific abstraction usually requires an agent to additionally reflect on her own behaviour. The same holds for concretion. An agent who concretizes swimming by particular movements usually does not reflect on the description of her movements – this only occurs during the respective learning process.

The fact that abstraction and concretion are spontaneous acts does not mean that there is no control of these acts but that the control is embodied. This leads to a reduced rational control by the agent with respect to abstraction and concretion. It does not mean that rational control of these acts is not possible but usually it only takes place in a subsidiary way. Often the reason is that abstraction and concretion do not appear as isolated acts but in connection with other acts subsumed into one focus.

Most human capacities are not innate but acquired by learning. Children learn abstractions from others mostly by imitation and not by explanation. Since generally abstraction and concretion are not analytical processes the respective learning is mostly implicit.[25] In particular the assumption that capacities are general and timeless is an abstraction itself since in reality capacities run through a continuous process of development. The more we use specific concepts and build houses the more our respective capacities develop. In this respect capacities are fundamentally dialectic; they are related to equivalence classes but at the same time undergoing continuous development.

As mentioned before, abstraction and concretion are not necessarily unknown to the agent but they are mostly executed in subsidiary awareness. If the agent focuses on the respective acts she can become completely aware of them and rationally trace them. However, usually this is too elaborate for the agent and therefore inefficient so that agents usually only switch to this mode if errors or problems occur. We will come to this point in the next section.

In the following we will investigate how abstraction, concretion and the respective capacities are related to action and knowledge as the main subject of this essay. However, without a detailed discussion of abstraction and concretion we are not able to understand the crucial aspects of both.

Knowledge and Action

While abstraction and concretion describe the elementary steps of thinking and acting, we now turn to the concept of *action* as the further developed mode of activity. We have seen that the elementary processes of abstraction and concretion are performed by the agent as acts in subsidiary awareness, i.e. they are usually not the focus of our attention in an activity. If we turn to the acts on which the focus of our attention is placed we come to action – in the following we always refer to intentional action, i.e. actions that are directed towards an objective that the agent is aware of. The objective must be abstract since it refers to a state in the future that the agent wants to achieve in analogy to similar situations in the past.

Consequently the agent starts the action with the determination of the objective and the situational preconditions by identifying the (abstract) objects which are supposed to take part in the process of execution. This requires the abstraction of the concrete interaction and dependencies of the involved objects; in this abstraction agents concentrate on their externality. To build a house we first have to identify the objects that we need for the construction, e.g. mortar and bricks. To this end we have to distinguish between the objects that are to be included in the process from those which are to be singled out. Ruben calls this the *classifying* activity of an action.

In the execution the agent turns from classification to the actual *processing* activity. This means that the externality of the involved objects is resolved and the concrete process (i.e. the execution) negates the initial abstraction. The objects become moments of the process. For example, in the actual construction process mortar and bricks do not appear in their individuality any longer but are grasped and built into the walls. In the unity of this process the objects become internal and inseparable, subsumed under the objective. At the end of the process the externality of objects is restored in altered form, e.g. as a new house that includes bricks and mortar. The (concrete) negation of objects in the process is (concretely) negated again and completes the sublation.

The entire action appears as a connection of abstractions and concretions that we describe in the following compilation:

(A1) *Classifying activity:* concrete environment → Action plan
(C1) *Processing activity*: action plan   → Action execution
(A2) *Evaluating activity*: action execution   → Outcome

The action plan describes the abstraction that relates the identified objects in the given situation with the intended objective of the action. The action plan describes how the agent intends to achieve the objective.[26] The execution consists in the concretion of this action plan through concrete execution. Every action is completed by the *evaluation* of the outcome, i.e. a comparison of outcome and objective of action. It is essential for an action that it is either successful or a failure depending on the evaluation. The evaluation is the precondition for the agent to learn from an action. By the evaluation the agent realizes a closed feedback loop, i.e. the agents can learn from their actions and improve their underlying capacities. For example, at the end of the construction of the house the constructor has to check whether the house actually fulfils the expectations and what might be mended.

However, actions do not only involve an action plan, an execution and an evaluation. They also involve the agent's *volition* to achieve the objective taking the action plan as guidance. This volition starts with the *decision* to act and ends with the completion or abandonment of the action. The volition is the central aspect that assures that the action is in the focus of the agent's attention.

In the same way as abstraction and concretion require respective capacities an action is also based on corresponding conditions of which the agent must be aware. The existence of an explicit action plan and the evaluation are expressions for this awareness. The action becomes transparent for the agent – in contrast to abstraction and concretion where this control can only take place in secondary processes – and the corresponding capacity becomes a *rational capacity* that is subject to the agent's reflection according to Kern.

As we said before, even abstraction and concretion can be performed as proper actions. For example, an abstraction can be based on a formal check of properties, e.g. does the object have legs and a board to be a table. In this case the agent starts with the individual properties and ends up with the sublation of these properties in the identification of the object. In the same way a concretion can be completed by the explicit check of the outcome to become an action. Such *extensions* of elementary acts are usually applied to deal with problems or errors.

Kern has argued that the rational capacity is identical to the *knowledge* required for the successful execution of a particular action. Knowledge as an acquired capacity is subject to a process of learning, i.e. the process of developing this knowledge. We even find a mutual dependency of knowledge and action. Action fosters the development of knowledge and knowledge is a precondition of learning. This seems to lead to a hen and egg problem and raises the question of how agents come to new knowledge. The answer to this question is twofold: (1) either the agents create this knowledge in action by emergence (implicit learning), or (2) the knowledge is transferred in an abstract way, e.g. via communication (explicit learning). Although both ways appear as quite different at first glance, they largely resemble each other. Polanyi's description of indwelling gives us an idea about the cause of this resemblance.

Let us first consider case (2) of explicit learning. Typically explicit learning is based on instructions or rules. Instructions enable agents to perform completely new actions that they have never executed before. The instructions separate the action into a set of subactions for which the agent must possess the respective capacities and which are strung together in an abstract external way. As Polanyi has pointed out, by repetition of the connected execution of the subactions the agent finally becomes independent of the abstract instruction so that the action is merged into one concrete execution. Polanyi has called this process of learning *indwelling*. He has asserted that by indwelling the awareness of the respective process becomes subsidiary to the agent. The outcome of indwelling is that the agent has acquired a new capacity which she can identify (and control) in an abstract and rational way, i.e. as knowledge. The acquisition of the underlying knowledge is to be considered as sublation of the original instruction. Information has been transformed into knowledge. Whereas the agent originally depended on the instructions (or rules), the indwelling leads to an internalization that makes the agent independent of explicit instruction. The rules are no longer constitutive for the action.[27]

If we come to the implicit learning mode the situation is not completely different. The agent starts with some activity for which she already possesses the rational capacity. By further actualization, however, the character of the capacity can change, i.e. it might show a tendency towards another quality. For example, an agent who was able to build small sheds might develop the capacity to build entire houses in the course of time which is definitely a new capacity. In this case sublation also appears in the learning process via indwelling.

One aspect of action that we can clearly recognize in the explicit learning mode is the hierarchical structure of action and (consequently also) knowledge. The instruction of action by a set of rules requires that the action can be decomposed into sub-activities that constitute the super-action. Indeed this hierarchical structure of action and its complexity are well known.[28] As we said, the agent is only subsidiarily aware of most of the sub-activities. Let us take the example of a ticket machine. In order to buy a ticket the agent must be able to read the instruction on the machine, select the right destination, read the price correctly, find the right coins in her purse (collecting coins that make up the required value), and so on. If an agent uses a ticket machine most of these acts are executed implicitly and the only focus is to buy a ticket that defines the action. All mentioned steps can also be executed as proper actions but usually the agent performs them more or less automatically. However, all these subsidiary acts have been learned with some effort, e.g. in school. The learning process is due to indwelling.

According to dialectical thinking the question may be posed how the concept of knowledge as rational capacity to act is related to the standard analysis of knowledge known from analytical philosophy. In the following we want to show that the concept of knowledge as rational capacity is a sublation of the standard analysis of knowledge, i.e. it covers the key aspects of the standard analysis and resolves certain problems of the standard analysis. To this end we start with the standard analysis (SA) of (propositional) knowledge which has been expressed as

$$K_{(s,t)}p = B_{(s,t)}p \& J_{(s,t)}p \& p \tag{SA}$$

Here $K_{(s,t)}p$, $B_{(s,t)}p$ and $J_{(s,t)}p$ mean that at time t the subject s knows, believes and can justify the proposition p, respectively. The capacity of an agent to identify a concrete object in an abstract way already corresponds to some knowledge. For example, if the agent can correctly identify apples then the agent must possess the knowledge of what an apple is and be able to use this knowledge in relevant situations. Regarding p as an abstract, this means that $K_{(s,t)}$ p requires s to be able to identify situations in which p can be concretized, either in practical treatment or theoretical consideration. For $K_{(s,t)}$ p this corresponds to the truth of p. Another precondition is that the agent must be aware of her capacity to identify p. This does not exactly correspond to $B_{(s,t)}$ p since the latter describes the agent's belief that p. However, it is hard to imagine that s believes that p but cannot imagine cases in which she is able to identify situation that correspond to p. Finally, $J_{(s,t)}p$ reflects the experience of s to possess the capacity to identify objects that correspond to p. Such a justification can be practical or theoretical. To summarize these findings we can say that the standard analysis (SA) describes the fact that s disposes of the abstract connection p. This abstract connection is actualized in the concrete processes of identifying p in an actual environment. The acts of identification correspond to the abstraction of p.

To come to an evaluation of (SA) we can say that the standard analysis completely ignores why knowledge is required at all. The answer given by the capacity conception of knowledge is that the agent is expected to apply her knowledge in successful action. In standard textbooks of analytical epistemology we find that action's role is not even explicitly mentioned with respect to knowledge. The reason is that knowledge is to be understood as an abstract state and not considered from a process point of view. The failure of traditional KM approaches tells us that the standard analysis is obviously insufficient and we have to refer to the concrete contradiction of this conception to arrive at the crucial shortcomings.

The more often an agent actualizes (makes use of) a concept the further the knowledge develops. The outcome of concrete processes guides this development. We know what a 'chair' is if we are able to identify a chair and use it. We can apply the same view to propositions as abstract connections. This means that an agent s knows that p if s is able to identify p as a concrete object (in a given situation). Let us take the example p = 'snow is white'. To know this s must at least know the concepts of 'snow', 'is' and 'white'. Let us further assume that s stands in a field covered by snow. In order to know that p, s must first identify 'snow' and 'white' in the concrete environment. Moreover, she must know what it means that 'p is q' where p is an object and q is a property. Finally she must possess the capacity to identify the abstract connection 'snow is white' in the concrete object of her perception that she is standing in a field of white snow. This identification actualizes her capacity related to the knowledge that p. This experience of standing in the snow-covered field will again influence her capacity since it influences her experience. For example, if it is the first time that she has seen real snow, the experience will definitely deepen her knowledge related to the abstraction 'snow is white'. Also here we find what Polanyi has described as indwelling.

To show the consequences of the sublation of knowledge and action we go to the problem of epistemic closure. Taking only the logical point of view to this problem leads to a completely different conclusion than the dialectic view. The problem consists in the question whether the following conclusion is true: If an agent S knows that p and knows that (p → q) then S knows that q.

From the logical point of view the principle of epistemic closure makes sense. From the dialectical point of view it is a false conclusion since it neglects the concrete act of learning, i.e. that S comes to the conclusion that q in a concrete process of thinking that takes place in time. While logic aims at truth and timeless external relations, dialectical thinking aims at effectiveness. In particular this example shows that knowledge cannot be completely understood without dialectics since it essentially relies on the process of its acquisition.

The standard analysis has led to some misjudgements, e.g. that knowledge consists of some resource that *is applied* to action without any change. In contrast, the current concepts of knowledge and action expose the contingency of concretion. We are well aware of the fact that every action can fail due to adverse circumstances. This is an essential characteristic of action. It is the reason why we can falsify knowledge by action. That the contingent character of action is not generally acknowledged can be realized, for example, if we refer to Hawley.[29] In her discussion of (practical) knowledge Hawley has stated that it is sufficient to restrict the context of action appropriately in order to obtain a causal relation between knowledge and successful action. Kern's as well as the current discussion show that such a causal relation can never be achieved.

Although Hawley's remark is correct that we can successfully perform actions which we have never executed before, only on the basis of instruction, she overlooks the fact that we must have performed at least similar actions in advance to be reliably successful. If Hawley gives the example of a life vest that we can take on in case of emergency even if we have never done this before, this scenario ignores the fact that we have performed other related activities, e.g. putting on backpacks, jackets, etc. This means that instruction can be helpful but we have to realize that the relation to experience beyond mere instruction is necessary.

In the final part of this essay we will discuss which practical conclusion for KM we can draw from the dialectical character of knowledge and action.

Practical Consequences

In the following we want to examine the practical consequences for knowledge management systems (KMS), which result from the described concept of knowledge and its dialectic relation to action. As explained in the introduction the experience with technical solutions in KM suggests that the traditional abstract-centred concept of knowledge is not suitable to tackle the challenges that we face in modern organizations which are characterized by continuous learning and innovation. In the previous section an analysis of the reasons for the failure of the abstract-centred concept of knowledge on the basis of philosophical arguments has been given. We have seen that the view that a person immediately acquires knowledge simply by the 'consumption' of information ignores the process of learning, the actualization of knowledge, and the process of indwelling, i.e. the specific process aspect of knowledge.

There are some results of the previous investigation which have to be considered for the design of KMS.

1. Knowledge is not 'contained' in an abstract knowledge artefact but these artefacts are involved in continuous processes of concretion and abstraction.

2. Knowledge is not a general and eternal state but a continuously developing rational capacity.

3. Knowledge becomes manifest in action and action is the source of knowledge.

Most theories of propositional knowledge in contemporary epistemology regard it as a relation between an agent and a true proposition, i.e. as an external connection.[30] The relation is mainly considered as a belief relation augmented by the agent's capacity to give some justification for it. It is clear that such a view suggests that it is sufficient to supply an agent with true information (proposition) from a trustworthy source (e.g. a KMS) that implicitly provides some justification and suffices to make the agent believe that the provided information is true in order to transfer knowledge to the agent.

Walsham has pointed at the implicit basis of shared capacities which are required to make such *simple* knowledge transfer work.[31] In his investigation he referred to Polanyi's concepts of sense-giving and sense-reading.[32] Sense-reading describes the process of making sense of a given knowledge artefact which can be seen as a first actualization. Sense-reading shows that KMS as they are designed today do not address the process character of knowledge that appears in sense-reading. The abstract representation decomposes the experience into several elementary expressions which the consumer of such a representation must transform back into action. This action reconstitutes the respective knowledge. The conclusion made after consideration of this process is that the knowledge representation should reflect the main structure of the respective action in order to support the transfer sufficiently.

To more efficiently realize knowledge transfer, KMS must better support the indwelling of actions based on instruction or other knowledge artefacts. Instruction-based knowledge is always shallow in the sense that the agent is often overstrained when a situation occurs that is not covered by the instruction. This is exactly what makes the difference between an expert and a novice in a certain domain. This requires a new perspective to knowledge artefacts. It is not primarily important to make the respective information available but also to adapt the provided information to the respective actions.

To this end, knowledge artefacts and the related actions are considered in an integrated way. First of all this requires an appropriate formal representation of action as a task object. The introduced task formalization contains the central formal characteristics of actions based on activity theory.[33] Activity theory takes the main structure of action into account by explicit inclusion of the agent, the objective/outcome, and the objects or tools involved in a process of action that is constituted by (1) agents, (2) objects and (3) objectives, which points to some outcome.

The touch point with the traditional form of KMS is the treatment of the included objects or knowledge artefacts. The new aspect is the extension to representations of the agents, the objectives and the outcome. The central problem in the approach of formalizing actions is that action representations are abstract and the fundamentally concrete character cannot be grasped by such representations. The rationale of why we nevertheless regard the approach as promising is that the action-based provision of information is much closer to the actual situation of the agent who uses the information to perform the action. In particular, it is possible to represent the history of a task including the contingent aspects of the respective action. This is for example done in task journals that describe task events and at which point of time knowledge artefacts were involved in the task.[34]

It is not only the representation of work activities in abstract form by which more efficient knowledge management can be obtained but also the form of interaction between users and KMS that can be improved. To this end the provision of knowledge artefacts is to be realized as an offer of services by which the KMS supports general work activities, not only task management (TM) activities. Here it should be noticed that traditional TM systems are designed as applications among other applications such as editors or business systems and ignore the fact that work activities become manifest in *all* interactions of users with applications on the desktop. This means that the TM system must be extended to a TM framework that supports activities in all desktop applications. However, if we realize that work activities appear in all applications that users employ in their interaction with computers then we see a necessity to include TM aspects in all kinds of work activities. Consequently, TM functionality must be included in all applications. Only in this way can we grasp the actions that form the relation to knowledge.[35]

To induce knowledge the KMS must not only include action support as TM functionality in all applications, but they must particularly *guide* the users in performing tasks. By this guidance users learn to better perform their task by direct support of execution which results in the acquisition of knowledge. Of course knowledge artefacts still play a central role in this process. However, they are mediated by the task structure. They are embedded in a work context which gives them a meaning. Moreover, this context is important later when other users want to understand and use these knowledge artefacts. It provides additional information about the knowledge artefact that gives better guidance with respect to the utilization of artefacts. In other words, the task better mediates action and knowledge artefacts.

The main difference to traditional KMS is that the knowledge artefacts are tied to the respective action as the central hub to knowledge acquisition. It is expected that in this way task information can be used to better transfer knowledge. Although tasks are abstract representations they are often too context-specific for general usage and transfer of knowledge. To increase the transferability of task information further abstraction is often required. The required transformation necessitates additional identification of individual abstracts which cannot be performed analytically and is not automatically executable. Here the user must be included to provide the necessary abstraction. To transform these abstractions into action again users have to concretize them again, e.g. by finding suitable resources that fit the general descriptions. These concretions must be supported by specific knowledge services.

It is not only the representation that must be changed but also the user interaction with the KMS. This concerns the way that knowledge artefacts are created and used. The traditional approach sees the creation of knowledge artefacts outside the actual work activity, i.e. based on reflection on the work done in the past. The idea behind this approach is that users have 'collected' knowledge that they afterwards cast into knowledge artefacts. However, if knowledge is mainly incorporated to concrete work activity and a considerable part of it is tacit, it is rather difficult to develop corresponding knowledge artefacts ex-post; at least it is impossible on an everyday basis. Representing the daily work activities and augmenting them by additional knowledge artefacts based on *actual* performance yields a more realistic picture of how the respective tasks have been performed. This particularly holds for information that otherwise would remain implicit and only becomes explicit by the recording of activities.[36] This has been considered in the design of the task-based KMS in order to optimize the adaptation to the way users interact with the system.[37]

It might be argued that the approach again focuses on knowledge representation, i.e. abstraction, rather than concretion. Above all, we have keep in mind that knowledge only becomes manifest in the users' concrete actions which take place outside the KMS. Nevertheless the presented approach addresses both abstraction and concretion: abstraction is extended in terms of task representation while the design of the KMS influences the users' concretions such as organizing their task, creating tasks during work activities, and thus implicitly explicating their knowledge incorporated in action. This explication does not give direct instruction but rather shows by example how to proceed. Nevertheless the question may be raised as to whether this proceeding actually solves the problems of focusing on abstractions in IT-based knowledge management. IT applications are generally characterized by a focus on symbolic representation and abstraction. We have to accept that we cannot completely overcome the principle separation of abstraction and concretion. However, we can identify concrete contradictions and sublate the conceptual framework. In this way we arrive at new abstractions that might better fit the requirements of the concrete actions under consideration. The better we understand the contradictions, the better are our opportunities to further develop the involved abstractions. The introduction of the explicit task model yields a sublation that provides a description that comes closer to the actual activity by including the work context.

The fact that experts can perform action intuitively and more proficiently than novices who have to plan their action carefully and stepwise is related to indwelling. This imposes a principle barrier in transferring knowledge from experts to novices since indwelling as implicit learning is difficult to support by KMS. The experts are often not even aware of their proceeding or omit individual steps that are necessary for novices who have not yet mastered the process as a whole. Therefore it is important to grasp the development of individual persons in executing specific tasks. Tracking the history of individual proceeding shows how the respective capacity has been acquired. In particular, it provides information for novices of where to start with their own proceeding.

Whether one abstraction is actually better than another is a question of experience that cannot be decided by philosophy. Philosophy can provide a problem analysis and point at hidden contradiction. The sublation must be left to the domain experts and is subject to validation by experience. In this respect dialectics is not a means to predict future technologies but a means to analyse the conceptual basis of present technologies and reveal hidden tendencies. This, however, is a crucial task since the best technology cannot succeed if the underlying concepts are not appropriate. A better conceptual framework opens up new opportunities and ways to cope with the world in which we live.

Discussion

In the current essay we have applied dialectics to develop the concept of knowledge as rational capacity to better deal with the dynamic aspects of actualization in action and learning, i.e. knowledge adaptation. In this respect we have pointed at the limitation of abstraction. A question that may be raised in this respect is why abstraction works in a world of contingent processes in which the distinction between objects is in principle fuzzy. However, if the world would be completely erratic, planned action would be impossible. Although it is far from obvious that we find the necessary stability required for abstraction, it is simply a matter of experience that abstraction is a means to successfully cope with the environment.

We can answer this question by pointing at the fact that the distinction between objects and processes is a relative one, depending on the timescale in which changes take place. A chemical reaction is a typical example. The reactant and the product of the chemical reaction generally show a much longer lifetime than the time required for the reaction. This is the reason why we consider the reaction as a process and the chemical materials as objects. These islands of stability are addresses by abstraction. Consequently the same object can appear as a process, when considered in a more stable environment, and as an object that is grasped by abstraction, when considered in a changing environment. A biological species is rather stable but in evolution we consider the process of changes in these species.

On the other hand, we may pose the question of whether the dependency of objects on agents' point of view does not inevitably lead to relativism. To answer this question we can point at the fact that concretion always refers to one reality as the unique standard. And it is a particular goal of dialectics to resolve concrete contradiction that might lead to inconsistencies. Sublation is the central means to resolve these contradictions and restore a unique representation. It is to be remarked that the presented position assumes that concrete contradictions are caused by the world but that the reason for their existence is the limitations of abstractions that we need as means to cope with the world. Dialectics does not contradict realism but must be seen as a means to ensure realism in order to resolve contradictions that originate from abstraction.

Traditional analytical philosophy strives for the same goal. However, it misconceives abstraction as an adequate way to completely describe the world. Thus analytic philosophy tries to resolve concrete contradictions that appear in language games by introduction of more complex abstractions and ignore that these contradictions result from the process aspects of reality. In the same way as medieval scholars tried to rescue the geocentric system by introducing new corrective parameters, modern analytical philosophers try to solve the occurring contradictions by introduction of more complex descriptions. The concepts of action and knowledge are typical examples of this approach. Craig described in an illustrative way how the introduction of new conditions solves particular problems of the standard analysis of knowledge but systematically leads to new problems. Some discussions in modern analytic philosophy resemble the argumentation in the case of the sorites paradox. In his recent article, Dudda supplements the standard analysis of knowledge in order to rebut the example given by Gettier.[38] To this end he introduces a new condition that refers to the reason r by which the subject justifies that the

$$K_{(s,t)}p = B_{(s,t)}\,p \& J_{(s,t)}\,(r,p)\; \& \;(r \to p) \tag{SA}$$

where $J_{(s,t)}$ (r,p) means that the subject s justifies p at time t by the reason r. However, the attempt fails since it only shifts the problem. If, for example, s has simply guessed r, then even if r as well as $r \to p$ and thus the justification are true, we would not count this a valid justification (from a logical point of view) since the justification for r is missing. The proceeding cannot work even if it looks correct at first from a logical point of view. The mistake is that it ignores the fundamentally dialectic nature of the problem.

Even if not mentioned here, language plays a central role in abstraction. However, as Davidson has pointed out, abstraction is not restricted to language and human beings but is something that all creatures – especially higher animals – use as they systematically interact with their environment.[39] It is the mastering of language that centrally distinguishes human from animal thinking. Language can be seen as a concretization of abstraction materialized in symbols. As such it allows for a more extensive usage of abstraction and particularly mathematics in its full complexity is difficult to image without language.

The development of abstracts finds a parallel in the history of language that reflects the process of human learning and environmental changes. Thus the development of abstracts reappears in the shift of meaning of words. Especially long-term developments such as those in language demonstrate that abstracts are not arbitrary or subjective. We observe social mechanisms that stabilize the short-term usage of language in order to ensure its adequacy in describing the environment. The speed of change generally reflects the changes in this environment; words that denote natural species are extremely stable while words that refer to temporary fashions tend to disappear rather quickly. Even if language is not a necessary precondition for abstraction it is the basis for the communicability and universality of abstractions.

Finally, it is to be remarked that the structure of action and knowledge and the underlying relations between process and object descriptions reflects the fundamental structure of the world in which we live. This leads to ontological questions that cannot be discussed here. Thus, we find an entwinement of law-like (abstraction) and contingent behaviour that resembles fractal structures as described in chaos theory.

14 ACTION-DIRECTING CONSTRUCTION OF REALITY IN PRODUCT CREATION USING SOCIAL SOFTWARE: EMPLOYING PHILOSOPHY TO SOLVE REAL-WORLD PROBLEMS

Kai Holzweißig and Jens Krüger

Problem: Action Direction in Product Creation

Operating in a global market that is characterized by high competition, growing customer demands and steadily shortening product life cycles, an efficient management of product creation processes plays a key role for manufacturers of complex products. According to Ohms, product creation processes encompass all activities prior to series production of a product, starting from the initial product conception, shifting over to product engineering activities, as well as planning of manufacturing equipment, supplier integration and final production ramp-up.[1] Two salient characteristics of product creation processes are (a) the high amount of division of labour involved and (b) their immense knowledge intensity.

In order to successfully keep a product project, e.g. the development of a new car series, on schedule and to securely carry the project to its agreed start of production, mechanisms that foster the coordinated collaboration of all actors within product creation are – due to the inherent high complexity of such projects – of high importance. A core instrument employed is so-called 'product creation process models', in which the coordination and temporal synchronization of the major process steps, as well as their causal dependencies, are formalized according to a stage gate approach.[2] Knowledge, such as that employed in these process models, is the basal precondition for action.[3] Process models serve as an important instrument in project management to plan, steer and report product creation projects. Even more interesting, for the aspects discussed in this essay, is the fact that process models are used by project members to derive, negotiate and execute action plans. Such action plans make use of the information available for each process step, namely *what* has to be done by *whom*, in *what time*, with *which inputs*, and *what are the expected outputs*. Hence, process elements and their contents carry a deontology in the Searlian sense, which puts the actors assigned to a certain process step under the obligation to perform as the process prescribes and which gives them also certain rights and powers to do so.[4] In this way, social reality in product creation is constructed. However, following Searle, this only works if three basic primitives are present, namely,

(1) collective intentionality,[5]

(2) assignment of function, and

(3) collective assignment of function.

Speaking in terms of process models this means that: (a) all project members must possess a shared intentionality regarding the project and its goal (doing something only as part of something); (b) they must be able to attribute function to what is described in the process model (attribution of meaning); and (c) they must do so in a collective fashion – this then gives rise to 'status functions' and thereby 'institutions'. Construction of reality fails if one of the three basic premises is not met, for example due to a lack of acceptance or attribution of function. Construction of reality, which is based on an agreed process model, serves in the timely direction of the actions of the different project participants. Consequently, if construction of reality fails (possibly through lacking collective intentionality), process models as a basis for derivation and negotiation of action plans become useless, which in turn yields negative consequences for project success.

However, an agreed process model of the product creation project to perform does not automatically result in the optimal or even expected construction of reality that serves as a basis for action direction. Moreover, it is the case – as constructivists[6] and empirical research[7] show – that individuals attribute different meanings to the entities in question and hence different individual realities are constructed. Such insights are contradictive to the positivistic model understanding, which according to Wyssusek[8] is prevalent in information systems and organization science.[9] Having explicated the importance of process models as a basis for directing action in product creation, we will now further address and explicate the inadequacy of the prevalent positivistic model understandings.

Analysis: The Prevalence and Inadequacy of Positivistic Model Understandings

Following Wyssusek, a positivistic world view rejects metaphysics of any kind, and does thereby not consider the ontological status of the epistemic objects in question.[10] Within the positivistic stance, only the sense experiences, on which all acts of cognition rest, are taken as given (positive) and (objectively) factual. Positivists account for intersubjective comparability of sensual cognitions through congruency of the human sensory apparatus. Hence, roughly speaking, the same sensory apparatus and the same sensory inputs yield the same cognitions. The role of presuppositions or predispositions of any kind in the active and 'creative' process of gaining knowledge is largely ignored by positivists.[11]

In order to explicate the outgrowth of the positivistic stance, let us consider the ongoing debate on the conception and nature of information. According to Kremberg, Wyssusek and Schwartz,[12] the prevalent positivistic position in information science views information as purely syntactical, a mere materialistic product (e.g. compare the general model of communication by Shannon and Weaver and the corresponding presuppositions discussed by Köck[13]). Regularly 'information' is confused, or even worse, equated with 'data', moreover sense and meaning is taken for granted (positive). Taking sense or meaning for granted is an identifying feature of a positivistic understanding – ontological questions regarding the constitution of sense and meaning, which is an inherent feature of information, are not raised.[14]

The positivistic understanding is quite prevalent in the domain of business process management, such as in product creation. Our practical day-to-day experience in modelling product creation process models shows that sense and meaning of such models is taken for granted as being straightforward by product creation actors. The view that process contents – such as descriptions or instructions – can be literally self-explanatory,[15] yielding the same attribution of meaning throughout all readers, is very common among managers and project participants. Such understandings of the very act of attribution of meaning stem from a mere reductionist view. Hence, it is no surprise that despite process models and detailed process descriptions, there is still a high amount of misunderstandings and conflicts in process work among product creation actors. Following a moderate constructivist approach, as well as evidence from empirical studies, we argue that attribution meaning and hence construction of reality differs from individual to individual.

In a study dealing with product innovation in large firms, Dougherty found that participants from different organizational units applied different meanings to the term 'market-oriented innovation'.[16] Dougherty argued that successful product innovation failed since the participants in the company were not able to 'speak one language',[17] thus bridging their different thought worlds – this is exactly what we find in our day-to-day experience. The importance of social context for the constitution of meaning and sense is also emphasized in works in the field of cognitive anthropology[18] (studies supporting cultural relativism), such as in research on emotion categories across languages.[19] Further positive evidence stems from empirical research on the weak version of the Whorfian hypothesis, namely that language influences perception and memory.[20] Evidence may also be found in the radical constructivist literature, such as in Foerster, although this stance is rejected in this essay.[21]

The socio-pragmatic constructivist approach by Kremberg et al. and Wys-susek explicates that epistemic knowledge, what is believed to be real and true, is always something subjectively constructed in our mind. These constructions are carried out against a background or previous body of knowledge,[22] which is shaped through processes of socialization. The continuous and reciprocal process of externalization, objectification and internalization, what Berger and Luckmann call the fundamental dialectic of society,[23] determines our horizon of interpretation and hence our horizon of meaning. In this way, constitution of sense and reality is always subjective and bound to social contexts and practices.[24] This is denied by the positivistic, as well as the solipsistic radical constructivist stance.

Returning to Searle's three primitives of human institutional reality, the insights above indicate that fulfilment of the second premise (attribution of function), as well as the third premise (collective attribution of function), is not straightforward. Given divergent attribution of function across actors with different backgrounds, a homogeneous collective attribution of function is hindered. In the following, we clarify the role and function of discursive practices, which we see as the key to mitigate the effect described above in order to create compatible constructions of reality.

Attributing Meaning to Models: A Two-Level Approach

As argued above, action plans in product creation that are subject to an efficient collaboration of actors are dependent on the underlying presuppositions in terms of shared or compatible constructions of reality.

According to Luhmann, the psychological system of an actor is operatively closed.[25] The constructed reality of an actor, which may be captured as a mental model,[26] is the result of linked thoughts (T), that are only accessible through interfaces (I), using compatible, reciprocal communications (C).[27] In this sense, actor X (e.g. the engineer) cannot approach the psychological system of actor Y (e.g. the after-sales person) directly. Moreover, it is not clear how the background of actor Y is composed, which would allow for inferences regarding possible construction of reality. An approximation of each other's construction of reality and attribution of meaning can only be achieved through discursive practices, which are embedded in suitable social contexts.

Following the Foucaultian tradition, Hardy, Lawrence and Grand define discourse as 'a set of interrelated texts that, along with the related practices of text production, dissemination, and reception, bring an object or idea into being'.[28] This is emphasized by Fairclough:

While I accept that both 'objects' and social subjects are shaped by discursive practices, I would wish to insist that these practices are constrained by the fact that they
inevitably take place within a constituted, material reality, with preconstituted
'objects' and preconstituted social subjects. The constitutive process of discourse
ought therefore be seen in terms of dialectic, in which the impact of discursive practice depends upon how it interacts with the preconstituted reality.[29]

Social reality, according to Hardy et al. and Fairclough, is constructed through
discursively constituted objects. This is also true for product creation process
models and their contents, which can be viewed as discursively constructed
objects. Discourse, as Hardy, Lawrence and Grand point out, does not solely
consist vis-à-vis conversations but can also include communicative practices via
emails, memos, telephone calls, intranet pages, etc (see Figure 14.1).[30]

Discursive practices are needed to elicit the presuppositions of actors and
to negotiate and execute action plans. In product creation, actors derive action
plans from a discursively constituted process model. The process model is a common objectification of the project contents for the actors and a central point
of reference for project work. According to the two-level model of process-oriented discourse presented in Figure 14.2, it must be distinguished between two
different levels of discursive practices in terms of (a) action-plan negotiation,
and (b) action-plan execution.

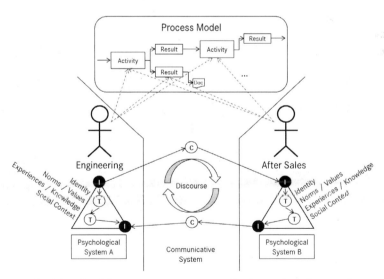

**Figure 14.1: The role of discourse in construction of reality and attribution of meaning
in product creation.**

At the operative empiristic level, actors carry out their project work according to the action plans derived – action plans are put to work. This is the level of day-to-day project work. If something goes wrong, e.g. expectations of actors prove wrong or misunderstandings about process contents occur, the participating actors move up into the model-oriented level of discourse. In the model-ori-ented level of discourse actors draw on existing discourses (the process model) and converse about their presuppositions in terms of those existing discourses, thus deriving and negotiating new or adapted action plans. Having clarified each other's constructions of reality and having hereby reached an agreement on how to proceed, the actors move back down to the operative empiristic level. In this way, such two-level discursive practices produce collective identity ('we-intentionality') which in turn may lead to collective action in terms of effective collaboration between product creation actors.[31]

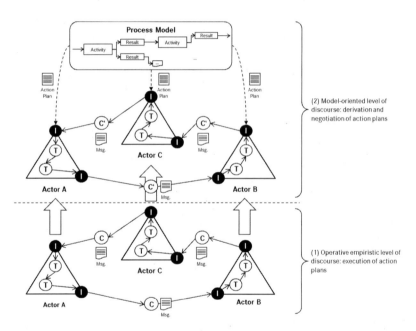

Figure 14.2: Two-level model of process-oriented discourse for action-plan derivation, negotiation and execution.

Implementation: Social-Software-Mediated Action-Plan Derivation and Negotiation

Given the two-level model proposed in the previous section, the question arises how actors in product creation can be adequately and efficiently supported in the derivation, negotiation and execution of action plans. How can IT tools provide support in mapping the social contexts and the existing discourses actors draw on in action-plan derivation, negotiation and execution according to our two-level model? We propose that a blend of different social software components linked to traditional business process and project management software meets the requirements described above. Komus and Wauch define social software as a type of software that focuses on collaboration and cooperation between actors.[32] The major characteristics of social software are its user-generated content, where generation of content is done by non-professionals. Social software is intuitive and easy to use, web-based and has a strong cooperative character. According to Komus and Wauch, social software can be viewed as comprehensive socio-technical systems. Popular examples are wikis such as Wikipedia, social network platforms like Facebook, and file sharing platforms such as YouTube. What all these platforms have in common is that they are supported through a community whose members contribute on a voluntary and self-organizing basis. Figure 14.3 shows a concept of a system that incorporates the insights discussed in this paper.

Pentzold points out that wikis can be thought of as an instrument to support the discursive generation of knowledge.[33] The inherent revision structures of wiki software yield a comprehensive transparency of the evolution of discourse. The complete history of discourse that actors draw on in text production, dissemination and reception is stored in a central repository that is accessible to all participants. Thus, explicit knowledge can be easily documented and put to use in terms of derivation of action plans according to our two-level model. Discursive mechanisms that support users in discussion, approval and commenting of action plans are available as well. However, it has to be noted, according to our explications in the previous sections, that the words written in such a wiki structure need to be annotated with the corresponding presuppositions of the author, knowing that these words are based again on certain presuppositions – an infinite regress.

Given that attribution of meaning and constitution of sense are always embedded in social contexts, wikis alone are insufficient. An instrument is needed that supports the socialization processes by creating ties between people that are a necessary prerequisite for the two-level discourse described above – ties for discourse in terms of action-plan negotiation and execution. This is supported through employment of social network structures and functions found on platforms such as Facebook. Implicit and explicit knowledge transfer is supported through social networks.[34] On platforms supporting social networks, actors keep an openly accessible profile on their person, which allows readers to draw inferences regarding the author's background.

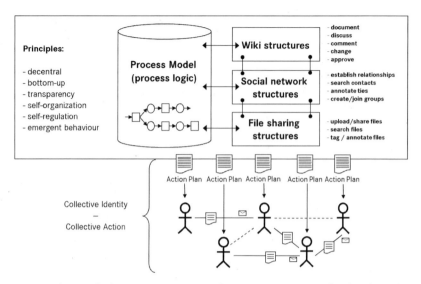

Figure 14.3: IT platform to support action-directing construction of reality through discourses.

Furthermore, actors can link themselves to each other by sending invitations, where links are annotated with information such as from which contexts the people know each other (e.g. process work, project work, etc.). Such annotated links might even contain information on the status of their relationships to each other in terms of successful action-plan negotiations in the past and possible problems that occurred. When there are questions and contents relevant to one specific topic, actors can create and join specific knowledge groups to foster exchange, as well as derivation and negotiation of an action plan. Given that a lot of explicit process knowledge is stored in existing files, such as office documents and drawings, the system employs file-sharing mechanisms. Such mechanisms offer file upload functions, as well as structured file search and download functions. Furthermore, actors can annotate and tag files using a collaborative folksonomy-based approach. All components and activities on the proposed platform refer to the product creation process model, which is stored in a business process management software to which all components connect.

Conclusion and Generalizations

In this essay we have taken a real-world problem from product creation and applied a problem-solving approach using insights and methods from philosophy. First, we analysed the problem by questioning the prevalent positivistic stance regarding the attribution of sense and meaning from both a philosophical and an empirical standpoint. Consequently, we argued for a new understanding in terms of a socio-pragmatic constructivist approach. In that approach, attribution of sense and meaning is always bound to social contexts and practices – socialization processes may create a common world reference, which determines the horizon of interpretation available. Based on these assumptions we introduced the role of discourse and discursive practices in creating compatible constructions of reality between actors, which fosters a better understanding regarding the active attributions of meaning. In a further analysis, we then introduced a two-level model for action-plan derivation, negotiation and execution, which employs such discursive practices. Finally, we discussed how the theoretical insights gained in this essay could be implemented by a collaborative software tool. Having implemented such a solution approach, there is of course the need to derive methods for measuring[35] its success in solving the initial problem. Here again, philosophical means can be employed to find the right measuring methods and in interpreting the results obtained.

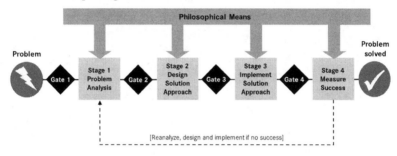

Figure 14.4: Generalized problem-solving approach employing philosophical means in all stages

The problem-solving approach employed in our example of product creation can be generalized, as shown in Figure 14.4. We propose that such an approach is useful to accommodate other complex problems in socio-technical systems as well. As we have shown in the example of product creation, we find philosophical means especially useful

- to understand the ontology of the entities in question (e.g. process models),
- to analyse the presuppositions of the agents in question (e.g. actors in product creation),
- to better understand the processes in question (e.g. action-plan derivation),
- to reveal category mistakes or possible fallacies we can run into (e.g. positivism), and

- to check the methodologies employed regarding their logical consistency (e.g. measuring methodology).

In this sense, philosophy has a high relevance in computer science, but also in other areas dealing with complex problems in socio-technical systems.

15 AN ACTION-THEORY-BASED TREATMENT OF TEMPORAL INDIVIDUALS

Tillmann Pross

Introduction

In this essay, I illustrate the relevance of basic philosophical research in information science by means of an example, where I present research on the semantic representation and model-theoretic evaluation of temporal individuals. While common ontologies underlying the computational treatment of temporal individuals are geared to the surface of natural language descriptions,[1] I propose to model temporal individuals by a constitutional ontology of distinctions among temporal variations based on the type of explanation which is used for the segmentation, identification and consequent representation of the respective temporal entity. With respect to explanations of temporal variation, recent investigations in action theory propose a threefold distinction between causal, behavioural and intentional explanation.[2] I adopt this to enrich the representational formalism of discourse representation theory[3] (DRT) with operators that specify how temporal processes are related to representations by means of explanatory identification of temporal individuals.

The Traditional Approach

Traditional approaches to temporal individuals usually follow Donald Davidson's logical analysis of action sentences,[4] where he proposed to capture the logical properties of natural language descriptions of actions with the introduction of a new class of ontological entity besides individuals, events, where events are supposed to be 'entities in the world with their own observer-independent grounds of existence'.[5] The following example illustrates Davidson's approach to the logical form of predicates that refer to actions.[6]

build a house: $\exists e. \exists x. \exists y. \, \text{agent}(x) \wedge \text{house}(y) \wedge \text{build}(e, x, y)$ (1)

While the Davidsonian analysis of reference to temporal entities seems to be acceptable at first glance, important information contained in the predicate 'build a house' is not represented in its logical form. First of all, this concerns Vendler's observation, who noticed that the temporal profile to which verbal descriptions refer differs for specific types of predicates with respect to their 'lexical-aspectual' class, in the case of 'build a house' that of an accomplishment.[7] It distinguishes the building of a house as the process of construction (i.e. the building) that brings about the house and its result that 'casts its shadow backward' in that it actually identifies the preceding activities as the building of a house. That is, the temporal profile of such a phrase goes beyond a simple event but is constituted by a more fine-grained substructure of processes, pre- and post-conditions. One can try to cope with this property of descriptions of temporal entities by establishing a substructure of events, as Moens and Steedman have proposed with their theory of 'event nucleus', where an event consists of a preparatory state, a culmination and a consequent state.[8] A very simple-minded approach can combine Vendler, Davidson and the theory of event nucleus within the framework of discourse representation theory as shown in the examples displayed in Figure 15.1:

build a house
$$
\boxed{\begin{array}{l} x, y, e \\ e : \text{build}(x, y) \\ \text{house}(y) \end{array}}
$$

Meaning Postulate 1:

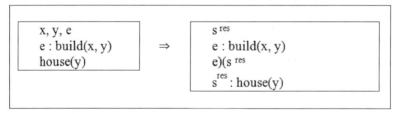

Meaning Postulate 2:

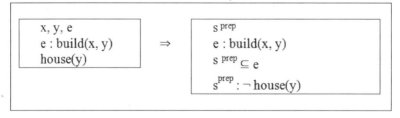

Figure 15.1: Lexical entry for 'build a house'.[9]

While this way of representing the semantic information contained in the example comes closer to the intuition about what 'build a house' actually means, there are still important problems to be solved.

First, probably the most obvious problem is associated with the adequate representation of the result of the event of building, i.e. that the house is supposed to come into existence if the process of building is properly finished. This is hard to capture within the standard framework of formal semantics because the condition s^{prep}: ¬house(y) in meaning postulate 2 does not capture the crucial point about a thing's coming into existence. It is not the case that the referent y is no house but that y does not exist at all at this preparatory stage of building.

It is doubtful whether the existence of a house is really a logical consequence of the building, namely a causal effect of the activities that make up the building, or if 'build a house' just makes a claim about the agent's intentions.

Basically, these problems have been tackled from two sides; syntactically with the introduction of additional predicates for 'staged' existence, becoming and causation[10] that specify the relation between the building and the house and semantically with a non-monotonic formulation of implication.[11] Both approaches have to face the fact that describing an action as 'build a house' neither logically implies nor causally forces the house to come into existence. Instead, 'build a house' intuitively describes the intention of the agent of building a house and it is this attitude of the agent towards the existence of the house that relates the activities of building to the existence of the house. In addition, it should be mentioned that the given preparatory and consequent states are not only distinctive for the building of a house, as there are other predicates that can describe the same constellation.[12]

Second, there is an ontological problem. Davidsonian events are supposed to be atomic model-theoretic entities such as individuals are. But the theory of event nucleus requires that events are split up in parts, which does not cohere with their fundamental ontological status.

Third, and this problem is closely related to the preceding one, the theory of event nucleus relies on a notion of state that has to be established at first (and preferably without reference to events to avoid a circular definition).

Fourth, Vendler classes can be coerced one into another, depending on the amount and type of information that specifies the temporal entity. That is, the Vendler classes are not distinct in the sense that there is a unique mapping between predicate and temporal profile.

At first sight these points may seem negligible and solvable by the good-will of the logician in charge of analysis, but they hint at a deeper problem of the traditional account of temporal entities that definitely appears when the model-theoretic treatment of representations of the above type is taken into consideration (see Definition 1). The discourse representation structure (DRS) itself does not provide information about the building of a house besides the trivial fact that it is an event of building.[13] As a matter of course, this information does not suffice to identify the building of a house. Perhaps the evaluation of DRS conditions for events can say more about the specific identity of 'build a house'?

Definition 1 *Evaluation of DRS event conditions (simplified)*[14]

Given a set of events and states EV structured by <, a Universe of individuals U and an Interpretation function I,

With respect to a Model M and a world w, g is a verifying emebdding of

$$R(x_1,...,x_n) \text{ iff} < g(e), g(x_1),..., g(x_n) > \in I(R)$$

Where g is an assignment that maps e onto an element of EV and x1,...,xn onto elements of U.

In simple terms, the DRS condition that represents the building of a house is satisfied in a model if the event referent e can be mapped to an event and the other discourse referents to individuals such that R can be embedded by the interpretation function I. Coming back to the above question about the identification of 'building a house', there is no additional information about what makes up the building of a house besides the trivial fact that it is 'true' if 'build a house' is contained in the model. That is, neither the DRS nor its evaluation conditions say something about what makes up the temporal profile of building a house that serves its identification. Instead, the semantics of an event is only concerned with the proper embedding of its arguments but not its temporal profile. There exists a final possibility that may help in solving this problem. In DRT, events are related to a time structure by means of a location function LOC that maps events to intervals of time. While this function seems to go in the direction of an answer to the question about the identification of events, the actual function of LOC has unfortunately never been spelled out in a way that it specifies the location of a given event. In addition, if actually spelled out, the function LOC would give a purely quantitative identification of the respective temporal entity. But the identification of 'build a house' goes beyond the statement of a certain amount of time, as the corresponding temporal entity is distinguished by its status as intention of the performer of the action. The loose ends of the traditional analysis of temporal entities entail further problems:

- Given a certain description, it is not possible to say something about why the event starts and ends, which in turn makes it difficult to justify a quantitative identification.
- As both events and states are equally mapped onto intervals, there are no criteria (and there is no need) to distinguish an event from a state besides their symbolic representation. This problem is eminently critical as the event nucleus relies on the distinction between events and states.

- The interpretation of temporal entities in the sense spelled out in Definition 1 is no interpretation in that it explains the entity, i.e. that it says something about what makes a set of intervals an entity besides the trivial fact that it is an entity.

All in all, this critical examination gives rise to the question of whether the traditional way of treating the reference to temporal profiles employs the right means at all. In other words, how should the temporal reference of a predicate be identified, if no 'essential and established facts' (refer to Searle[15]), i.e. an explanation of the identification about the entity in question is available? In addition, the many-to-one relation between predicates and temporal profiles makes it difficult to develop such a theory of explanatory identification from the surface of natural language. Instead, we should seek to develop a theory of how descriptions relate to temporal profiles based on a theory of how temporal profiles and consequently temporal entities are constituted at all; this is what the second part of this essay is about.

The Action-Theory-Based Account for Temporal Individuals

Recent psychological experiments suggest that, given a certain perception of temporal variation, humans structure the perceived temporal variation along the lines of 'goal relationships and causal structures'.[16] Consequently, the temporal entities resulting from event segmentation are to be understood as structures of temporal profiling imposed on perceptions and consequent projections of temporal variation.[17] The psychological insight that it is structured sequences of action that allow for the segmentation of temporal variation and that these structures are present in mind when segmenting events accords with one of the fundamental assumptions of DRT, namely that humans make use of mental representations (in particular when interpreting utterances). We can thus establish a natural relation between DRT and the psychological theory of event perception structures if we introduce plan-goal and causal structures as mental entities of representation. Before I proceed in spelling out how the fusion of DRT and the theory of event segmentation may be established, something more has to be said on the structures of temporal variation. The question of how humans explain temporal variation is one of the major topics in philosophy, especially action theory. While classical approaches focus on causal[18] or rational[19] explanations, more recent investigations propose a threefold distinction of explanations:

- the physical, design and intentional stance[20]
- the varying ability to have (meaningful) mental representations[21] in machines, animals and humans
- the culturally founded discrimination of movements from behaviour and intentional actions.[22]

Leaving issues of notation aside, all these approaches to the explanation of temporal variation have in common that they distinguish between three types of explanation: causal physical movement, behaviour (in its literal sense as goal-directed action triggered by desires) and intentional action (in the sense that it is rationally controlled behaviour). Structurally, these types of explanation are interrelated as both behavioural and intentional actions make use of the fundamental principles of causality to achieve their goals with respect to intended ends. Behaviour and intentions differ in that behaviour refers to a sequence of actions under the control of the agent that serve the realization of a goal triggered by a certain desire whereas intentions include an additional involvement of rational choice and commitment.[23] If we apply these considerations to the psychological insight that temporal entities are segmented along the lines of causal and planning structures, we should make use of all three types of explanations to extract entities from a given temporal variation. In particular, a temporal entity such as the present building of a house is to be segmented with the help of an intentional explanation.

Given these preliminary thoughts on the explanation of temporal variation and resulting temporal entities, we can now come back to specifying how this can be captured in the framework of DRT. For reasons of space, I will keep this as simple as possible. First, I assume that temporal variation is captured by a set-theoretic model structure of timepoints, i.e. temporal variation is modelled by a tree-like structure of possible times. Formally, such a structure can be achieved along the lines of branching time logic with respect to CTL*.[24] I omit the formal details here and only give a rough sketch of what the model is supposed to look like.

With respect to the structure of temporal variation, the model should contain a set of times T, where each $t \in T$ is annotated with the states of affairs that hold at the respective time. In addition, T is partially ordered by $<$ such that $<$ is allowed to branch. In a first step, we should then determine how to relate this structure to representations of temporal entities. Second, we can then examine how representations of temporal entities relate to natural language descriptions of these temporal entities and refer to specific profiles of temporal variation. We can interpret the model structure such that it serves the proposed theory of temporal entity extraction in terms of causal, plan-goal and intentional structures as follows:

- Transitions between times constitute the smallest units of causality, i.e. the atomic units of temporal variation.[25]
- Sequences of atomic transitions constitute a path. I assume for the sake of simplicity that plans correspond to such paths, where the final state of the path corresponds to the plan's goal.[26]
- Intentions are formed by distinguished sets of plans adopted by the agent.[27]

I introduce new DRS conditions that allow one to refer to specific constellations of temporal variation:

Definition 2 *DRS conditions for temporal entities*

- *Causality: If K is a DRS, x a discourse referent, then xCAUSEK is a condition*
- *Desires: If K is a DRS, x a discourse referent, then xDOK is a condition*
- *Intentions: If K is a DRS, x a discourse referent, then xINTK is a condition*

The crucial point is now to connect the syntactic representation with the model in terms of semantic evaluation. Several ways exist to formulate a semantics for DRS conditions as given in Definition 2. Again, I adopt a simplistic approach, where I make use of a class of assignment functions that assign causal processes (CAUSE), plans (DO) and intentions (INT) to agents at a certain time and restrict the requirement that the agent indeed has these attitudes towards her activities at times (and not intervals).

- xCAUSEK is satisfied at $t0$ if there exists a path from $t0$ to tn s.th. K is true at *tn*.
- xDOK is satisfied at $t0$ if there exists a path from $t0$ to tn such that K is true at *tn* but not at $t0,...,tn-1$ and K is among the agent's desires at $t0$.
- xINTK is satisfied at $t0$ if there exists a path from $t0$ to tn among the agent's intentions at $t0$ such that K is true at *tn*.

Of course, the information contained in representations as pictured in Definition 2 does not suffice to identify the temporal profile of a predicate. For this task, it is not enough to specify the type of explanation which is employed to identify the temporal profile in question, but as these explanations correspond to causal, plan-goal or intentional structures, these structures must also be applied when identifying the temporal profile in question. That is, time individuals consist of a semantic representation and a pragmatic profile in terms of a branching-time substructure as given in Definition 3.

Definition 3 *The lexical structure of predicates*

The lexical structure of a predicate consists of a semantic (SEM) and pragmatic (PRG) part, where OP *is one of the operators* CAUSE, DO, INT:

SEM,

$$\boxed{\begin{array}{c} <e,\ x\text{OP}K> \\ e = name \end{array}}$$

PRG specifies the identification conditions for a name in terms of
- *a path if it is a causal structure*
- *a plan if it is a plan or intention*

At this point I have to mention the interactions of explanations and tense I have omitted due to reasons of space, as it is of course only in the light of the constant progression of time and action that the use of the 'segmentation operators' CAUSE, DO, INT make sense. For example, a present intention that reaches into the future has an uncertain outcome, but this changes once the intention is accomplished – the intention operator is dropped and the representation of the respective predicate must be replaced with an updated one that captures the now realized sequence of actions and their real outcome in terms of causation. Finally, I can apply the developed machinery to the introductory example of 'build a house' for the case of a description of an ongoing present action as in the example described by Figure 15.2.

Summary and Outlook

Based on a critical examination of the traditional account in the context of descriptions of temporal entities referring to temporal variation, I have proposed a framework that bypasses the problems of the traditional analysis by reference to the psychology and philosophy of temporal segmentation which allows for grounding the evaluation and analysis of descriptions of temporal entities in causal, goal-directed and intentional structures of temporal entity segmentation. With respect to the importance of philosophy in information science, this essay demonstrated the usefulness of philosophy in information science not only with respect to the delineation of the larger picture of information science but also and in particular for the solution of concrete technical problems.

'I am building a house'.

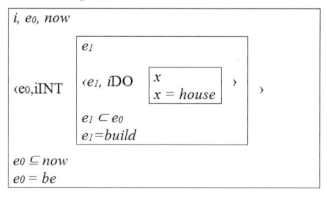

Figure 15.2: Example of an ongoing present action.

The proposed theory of temporal entities can shed new light on the meaningful processing of natural language predicates, the connection between planning, reasoning and representations as well as temporal ontology and knowledge management. Practically, probably the most promising application of the proposed analysis concerns the implementation in a robotic setup in the framework of goal-directed human–machine collaboration.[28]

16 FOUR RULES FOR CLASSIFYING SOCIAL ENTITIES

Ludger Jansen

Introducing Social Entities

Many top-level ontologies like Basic Formal Ontology (BFO) have been developed as a framework for ontologies in the natural sciences.[1] The aim of the present essay is to extend the account of BFO to a very special layer of reality, the world of social entities. While natural entities like bacteria, thunderstorms or temperatures exist independently from human action and thought, social entities like countries, hospitals or money come into being only through human collective intentions and collective actions. Recently, the regional ontology of the social world has attracted considerable research interest in philosophy – witness, e.g., the pioneering work by Gilbert,[2] Tuomela[3] and Searle.[4] There is a considerable class of phenomena that require the participation of more than one human agent: nobody can tango alone, play tennis against oneself, or set up a parliamentary democracy for oneself.

Through cooperation and coordination of their wills and actions, agents can act together – they can perform social actions and group actions. An important kind of social action is the establishment of an institution (e.g. a hospital, a research agency or a marriage) through mutual promise or (social) contract. Another important kind of social action is the imposition of a social status on certain entities. For example, a society can impose the status of being a 20 Euro note on certain pieces of paper or the status of being an approved medication to a certain chemical substance. Other social entities come along without a physical 'bearer entity', like electronic money:[5] the numbers on my account statement do not count as the money in my bank account, but are only signs for it. Such bearer-less social entities are established through the transfer of certain rights or obligations from, say, my employer to my bank where I, in turn, can claim them. Similarly, a credit card itself does not count as money (for were it to count as such, what value would it have?), nor does it represent any money (for how much would it represent?). Rather, using a credit card transfers the right to claim a certain amount of money from the credit card company, which will, in turn, claim it from me. Analogously, a health insurance card is a sign that a hospital will be licensed to claim treatment costs from the insurance company. In this way, the world is replete with a plethora of social entities which are highly important for our everyday life, for economics, politics and culture – thus, also for respective information technologies and their applications in these fields.

In this essay, I will discuss an application from medical information science – more specifically, the NCI Thesaurus. I will first discuss some ontological shortcomings of the representation of social entities in this thesaurus in the next section, below, and will then suggest four rules for classifying social entities (pp. 193–9), including the use of standard top-level categories, the characterization of specific social categories and the representation of ontological relations between social entities. Finally, I develop a small fragment of a social ontology to show how these rules can be put to work (Table 16.3).

Real-World Examples from a Medical Thesaurus

Though social entities as such are not governed by laws of nature, they are, as I will demonstrate, nevertheless important in areas such as medical information science. To confirm this relevance, I will draw on real-world examples from the National Cancer Institute Thesaurus (NCIT),[6] a terminology database designed especially for the needs of the US National Cancer Institute (NCI).[7] One function of the NCIT is 'the provision of a well-designed ontology covering cancer science'.[8] To achieve these goals, the NCIT aims to provide a 'true *is_a* taxonomic structure, polyhierarchy, inferred partonomy and other features that make it suitable for supporting complex query operations against appropriately coded data repositories'(as the NCIT describes itself in its entry 'NCI Thesaurus'). I will show, first, that as far as social entities are concerned, the NCIT is far from meeting these objectives[9] and, second, that the currently emerging philosophical (sub)discipline of social ontology can come to its aid.

Compared to other areas like genes or cell structures, references to social entities may be somewhat peripheral in the NCIT (see the figures given' in Fragoso et al.).[10] But they are, nevertheless, important, because the medical world is not disconnected from the social world. Cancer, the central topic of the NCIT, has social causes and effects. As Graham Colditz, the general editor of the *Encyclopedia of Cancer and Society*, puts it: 'Not only do health care providers and regulatory approaches each have a role, but individual behavior changes can substantially reduce the burden of cancer in our society'.[11] This is why the NCIT contains items like 'Stress and coping' or 'Social aspects of cancer'. Patients are never isolated individuals. Even lonely patients live in a social world and this social world may support or inhibit a healing process. This is why the NCIT contains items like 'Family', 'Minority', 'Support system' and so on. Moreover, healing patients is a business, governed by health policies and health administration. This is why the NCIT contains items like 'Business rules', 'Accounting', and similar things, as well as cancer research, which is a social activity in and of itself. This is why the NCIT contains items like 'Cancer study', 'Control group', 'Scientist' and 'Funding'. Historically, the NCIT actually started 'with a collection of local terminologies in use for coding documents related to managing science – funded grants, reports and intramural science projects'.[12] These aspects of relevance are reflected in the choice of social items which are contained in the NCIT. I will now review some of the ontological shortcomings in the representation of social entities in the NCIT.

(1) That the social world is not at the core of cancer research is reflected in the eclectic and unsystematic way in which social entities are selected for and ordered within the NCIT. Moreover, the NCIT has a clear national bias: it focuses on topics relevant for the United States of America. Thus, for example, the only item listed under the heading 'Underrepresented Minority' is 'American Indian or Alaska Native'. From a global perspective, there are, presumably, many more 'minority groups presently underrepresented in biomedical and behavioural research' (which is the NCI-definition of 'Underrepresented Minority'). Eclecticism concerns the things that are to be represented. Here, a choice of entities is understandable because NCIT targets only a specific topic, i.e. cancer research, which is necessary if the size of the database is not to exceed a certain limit. It is, however, a clear hindrance to data integration across national borders or topical domains.

(2) Furthermore, the NCIT is rather parochial in its horizon. For example, it simply regards 'Clinical Study' as being synonymous with 'Study'. But, of course, not all studies are clinical, some are, for example, literature surveys. 'Underrepresented Minority' is simply defined as a group 'underrepresented in cancer research'. But, of course, a group can be underrepresented in many other ways, too. 'Funding' is subsumed under the semantic type 'Governmental or Regulatory Activity'. But is it essential that funding is done by the government? What about companies, charities or endowments? While eclecticism is about the choice of entities to be represented, parochialism concerns the definitions given for the chosen entities. If the definitions are given as if such entities do not exist outside the topical field, i.e. outside of cancer research, this affects the interoperability with other terminology databases. It should be clear that representations of social entities are especially prone to parochialism, though it might also occur with representations of natural entities.

(3) Often the NCIT follows topical associations rather than ontological guidelines, as they are provided by, say, BFO or OntoClean.[13] For example, the item 'Business Rules' has thirty-nine sub-items. An NCI business rule is, for example, the 'Improve access' rule: 'Support the effective dissemination, communication, and utilization of HIV/AIDS information to all constituent communities of the

NIH'. It should be noted that this is a particular rule, not a rule-type. However, an ontology should deal with types only. Instances (like particular rules) are not related via the subsumption relation *is_a* to anything. Hence, this is not an example for the 'true *is_a* hierarchy' the NCIT intends to provide. Moreover, the quotation is a formulation of the rule, which cannot be used as a description of it: the rule is in the imperative mood, descriptions and definitions are in the indicative mood. The NCIT should, thus, affix a phrase like 'The rule that prescribes to' to the formulation of the rule in order to get a description of it. Nor is the plural of the term 'Business Rules' appropriate if the term is to feature in a subsumption relation. Rather, the following would be appropriate (I follow the typographical conventions of the OBO relational ontology as described in Smith et al.[14] and Schwarz/Smith[15] and use, for example, italics for terms referring to universals and relations between them and bold type for relations involving particulars):

Business_Rule is_a Rule

Improve_access **instance_of** *Business_Rule*

But most of the sub-items of 'Business Rules' are neither rule-types nor rule-instances. One sub-item of 'Business Rules', for example, is 'Employment Opportunities', which are (again rather parochially) defined as 'Jobs available at NCI'. These are, of course, no business rules at all. Rather, the process of filling employment opportunities with suitable candidates is something that is governed by business rules. Similar things are to be said with regard to 'Academia', 'Animal Sources', 'Business Commerce Fiscal Consultation', 'Commercial Sources', 'Completion Status', 'Contingency Fund', 'Contracting', 'Discipline', and so on. Even the 'FDA Modernization Act of 1997' is not a business rule, but a legal document that may contain such rules. 'Enhancing Accessibility to Health Care' is not a rule either. It is an activity, even if it is an activity that is prescribed by a rule. The monstrous 'Non-programmatically Aligned Cancer Center Research Member Section of Cancer Center Support Grant Application' is not a rule either. It is part of a document, i.e. of a support grant application written by a cancer centre, listing people able to support a certain research programme. The rule is to include such a section in an application. But this does not make the section a rule itself, nor is it a good reason to list it under the heading 'Business rules'. Thus, most of the items among the sub-items of 'Business Rules' are not business rules at all, but either documents containing such rules, fields governed by such rules or entities that such rules refer to.

Another example is provided by the sub-items of 'Funding': most of the purported sub-types are not kinds of funding but concern things somehow related to funding, like 'Concept review', 'Funding Category', 'Funding Opportunity' or 'Special Exceptions Process'.

(4) The NCIT often mixes ontological categories. For example, 'Clinical Research' is given as a synonym for 'Clinical Study'. But this cannot be true: clinical research is the overall activity which relates to a clinical study as uncountable stuff does to countable things.[16] The NCIT also confuses the 'Personal Medical History' of a person with its record, since the former is defined as a 'record of a patient's background regarding health and the occurrence of disease events of the individual'. 'Funding' is defined in the NCIT as 'a sum of money or other resources set aside for a specific purpose'. The NCIT takes this definition from the *American Heritage Dictionary*, where it is not the explanation of 'funding', but of 'funds', while 'funding' refers to the activity of allocating money or resources for a certain project.[17] In any case, it is incoherent to subsume something defined as a sum of money under the semantic type of 'Governmental or Regulatory Activity' at the same time, as the NCIT does.

(5) The NCIT entries often do not reflect actual properties of the entities classified. An item like 'Other Minority' cannot, of course, refer to the property of otherness instantiated by certain minorities. It makes sense only in relation to the rest of the classification given in the NCIT. It does, thus, not reflect the ontic structure of the social world, but only the classification in this database, which could just as well have been different from what it is. Such 'other' items (there are more than a hundred of them in the NCIT) are, of course, a means to secure exhaustivity of the classification on that level. The price to be paid for this is that the 'property' used for classifying comes into existence through the very classification in which it is referred to. There are several modifiers in the NCIT that are similarly troublesome because they employ epistemic notions for defining ontic matters, among which are 'None or Not Applicable', 'Not Defined', 'Not Otherwise Specified', 'Not Stated' or 'Unknown'.[18]

Four Rules for Classifying Social Entities

If a sound underlying ontology is necessary for a coherent terminology, a sound social ontology is necessary for a classification of social entities. In what follows, I want to suggest four ontological guidelines which should in the future, in a favourable course of events, be integrated in a unified ontology of the social world: (1) do not forget standard ontology; (2) classify a social *F* as an *F*; (3) respect specific social categories and (4) make explicit the ontological relations between social entities.

First Rule: Do Not Forget Standard Ontology

Most formal ontological dichotomies apply to the social realm, too. In particular, this is true of the three fundamental dichotomies that also lie at the core of BFO, i.e. continuant vs occurrent, independent vs dependent, and individual vs universal. While I am a continuant being wholly present with all my spatial parts at every moment of my existence, my life is not wholly present at any point in time: it has a lot of temporal parts and stretches out in time. I am also a bearer of many properties like my weight or the colour of my skin, the change of which I can easily survive, whereas these properties cannot change their bearer: they are dependent entities, while I am – ontologically speaking – an independent entity. Taken together, these two dichotomies yield three main top-level categories:[19]

- independent continuants (including Aristotelian substances),
- dependent continuants (including properties and relations),
- occurrents (including processes, actions and events).

Characterizations of these categories can be found in Table 16.1 where I also (for use in this essay) distinguish between properties and relations as two kinds of dependent continuants.

The third basic ontological dichotomy is the distinction between universals (or types or classes) on the one hand and particulars (or tokens or elements) on the other hand. All of myself, my life and the colour of my skin are particular entities. These particulars instantiate the universals *Human Person, Life* and *Colour*, respectively, and are, in their turn, to be subsumed under the top-level categories *independent continuant, occurrent* and *dependent continuant*.

Table 16.1: Top-level categories[20]

Term	Definition	Examples
Continuant	an entity that exists in full at any time in which it exists at all, that has no temporal parts and persists through time while maintaining its identity	
Independent Continuant	a continuant entity in which other entities may inhere, but that cannot itself inhere in anything	an organism, a heart, a chair, a symphony orchestra
Dependent Continuant	a continuant entity that inheres in or is born via other entities	
Property	a dependent continuant which is dependent on exactly one bearer	the colour of a tomato, the disposition of fish to decay, the function of the heart to pump blood
Relation	a dependent continuant which is dependent on more than one bearer	being larger than something else, being the oldest person in the room
Occurrent	an entity that has temporal parts and happens, unfolds or develops over time	the life of an organism, a surgical operation, a day, a concert

Second Rule: Social Fs are Fs

There are social entities that are continuants, like the American president or a national border, and there are social entities that are occurrents, like the inauguration of the president or immigrating into another state. Social entities, too, come along as particulars and universals, like the business rules in the NCIT. Moreover, social entities can also be classified according to the Aristotelian categories like natural entities: there are social quantities like prices, social qualities like academic degrees, social relations like being married to someone, social actions like a promise or a political manifestation, and even social substances or at least quasi-substances like companies or states.[21]

In the history of ontology, it has been a matter of dispute whether the same categories apply to natural and social entities[22] and, indeed, it can be asked whether the appropriate category of, say, *academic degree* is *quality* or rather something like *social quality*, which is then subsumed under a top-level entry *social entity*, disconnecting the classification of social entities from the classification of other entities. To answer this question, a reflection on the logic of the qualifying phrase 'social' is of help. There are some qualifying phrases like 'pseudo-' or 'bogus' which are alienating phrases. If you have something written by pseudo-Aristotle, it is not written by Aristotle. If you have a bogus proof, it is, in fact, not a proof. Other qualifying phrases, however, are separable, like 'living' or 'good'. If something is a living horse, it is both a horse and living and if someone is a good thief, he is a thief (though probably not good in an unqualified sense). I take it that the attributive adjective 'social' is not an alienating modifier but rather a separable phrase. If something is a social phenomenon, it is a phenomenon: if something is a social activity, it is an activity. Thus, as a rule, a social *F* is an *F*, *F* being some universal. This should be reflected in an ontology-based taxonomy: social activities should be classified as activities, social properties as properties, social circumstances as circumstances, etc.

As can be seen in Table 16.2, there are examples for all three of the top-level classes and for some of the Aristotelian categories among the social items represented in the NCIT. All of the sample terms in Table 16.2 actually feature in the NCIT, though under different top-level categories. Again, I follow the convention to use italics for names of universals or classes and the normal font for names of particulars.

Table 16.2: Social entities from the NCIT

Category	Examples
Independent continuants	
Individual institutions	American Cancer Society, University of California at Santa Cruz, United States, National Cancer Institute, ...
Universals of institutions	Company, Family, Hospital, State, University, ...
Universals of collectives	*Group, Minority Group, Research Personnel, Staff, ...*
Universals of role holders	*Employee, Employer, Scientist, Statistician, Laboratory Technologist, ...*
Universals of concrete social relatives	*Legal Relative, Participant, Responsible Person, Spouse, ...*
Dependent continuants	
Properties	*Academic Degree, Board Eligibility, Education Level, Employment, Marital Status, ...*
Relations	*Affiliation, Consent, Ownership, ...*
Occurrents	*Administration, Admission, Advising, Conference, Research, Submission, ...*

Third Rule: Respect Specific Social Categories

In addition to the general categories of formal ontology, there are categories specifically pertinent to the social world. A famous example of a label for a specific social category is 'institution'. Unfortunately, this label is ambiguous in natural language and is, in fact, used for three distinct, though interrelated, categories of social entities:[23] (a) For institutional rules (e.g. constitutive rules of the 'counts-as' type),[24] (b) for things instituted and (c) for the act of instituting something. Note that institutional rules (a) are to be categorized as dependent continuants while institution acts (c) are to be categorized as occurrents. In its own definition of 'Institution', the NCIT opts for a sub-type of institutional entities (b), defining 'institution' as '[a]n established society, corporation, foundation or other organization founded and united for a specific purpose, e.g. for health-related research'– the additional information 'also used to refer to a building or buildings occupied or used by such organization' should, rather, form a distinct entry. Thus defined, institutions are independent continuants. Examples are listed in Table 16.2, both for individual institutions and universals of institutions. But institutions in the sense of (b) can also be found among dependent continuants, like academic degrees or a marital status, and among occurrents, like rituals or festivals.[25] Furthermore, the three meanings of 'institution' have to be kept apart, which does not always happen in the NCIT. For example, 'marriage' can denote (a) the abstract institution of marriage to be found in some societies, but not in others, (b) a particular marriage, i.e. a couple's being married, and (c) the act of getting married. The definition of 'marriage' in the NCIT illegitimately confounds (a) and (b).

A specific problem of social ontology is also to differentiate different kinds of groups and of group unity[26] and to analyse social actions.[27] Various examples for groups represented in the NCIT can be found in Table 16.2.

An important variety of independent social continuants in Table 16.2 are the universals for role-holders like *Statistician*, defined in the NCIT as a 'person responsible for the compilation, organization, and analysis of mathematical data'. This definition closely resembles Bolzano's schema for the content of concrete ideas, i.e. 'something that has property P'.[28] Bolzano's concrete ideas are, of course, closely related to the corresponding abstract ideas of 'having the property P'. Similarly with (social and other) roles: roles as such are dependent continuants, but the thing that plays that role is normally an independent continuant. The social roles of employer and employee, of statistician and laboratory technologist are all played by human persons (who are then the role players). Though it is typical for humans to play social roles, it is neither essential nor typical for humans to play any one particular role. Playing a role is something contingent as opposed to having a certain function, which is something typical or even essential.[29] We can thus distinguish between the role player, the role and the role holder (a term also used, for example, by Mizoguchi et al.).[30] In our example, a human person as the role player and the role of being responsible for the statistics together yield the statistician as the role holder: or, to use another example, *Person* is not the parent of *Chairperson* (as in the NCIT), rather, persons are potential role players for the role of acting as chairperson.

A similar analysis applies to the concrete relatives in Table 16.2, which have the structure of 'something that stands in a relation R to some other thing'. A spouse is someone who stands in the relation 'being married to' to someone else. Here, too, we deal with a combination of a dependent continuant (in this case a relation) and an independent continuant. Because of their similar treatment, such relatives are sometimes called 'abstract roles',[31] and since they involve an independent continuant, they are themselves classified as independent continuants. Indeed, spouses can have properties like weight or age and cannot themselves be the property of anything – and it is the defining feature of independent continuants in BFO that they 'are the bearers of qualities and realizables; entities in which other entities inhere and which themselves cannot inhere in anything'.[32] On closer analysis, however, concrete relatives could be decomposed into a relation and its bearer. One reason for the comparatively small number of examples for social relations in the NCIT (see Table 16.2) is that many social relations are hidden in the entries for concrete social relatives, like the relation of 'is married to' is hidden as a constitutional part of *Spouse*.

Fourth Rule: Make Explicit Ontological Relations between Social Entities

Social entities are not isolated. They do not stand alone in the social world, but are interconnected to other social entities as well as to natural entities. Formal relations that apply to other realms of reality also apply to the social reality, like the relations 'is a', 'part of', 'depends on' and so on.[33] The NCIT, for example, contains both 'Social work' and 'Social Worker', but they remain completely unconnected. But of course, a social worker is someone who is trained or employed to do social work; this is the expected role performance of a social worker. A necessary precondition for making this explicit is to list all entities referred to in the definition themselves as items in the ontology.[34]

A specific social relation is the relation of membership. While some have argued that membership is just a variety of parthood,[35] there are good reasons for regarding it as a (social) relation in its own right. Parthood is a transitive relation, membership is not; the same members can constitute several distinct groups, while the same parts uniquely assemble to exactly one whole; localized parts form localized wholes while localized members can form non-localized organizations.[36]

Social relations can be at times quite complicated and are not at all well accounted for by the subsumption relation alone. For example, the 'Subcommittee B Basic Sciences' of the National Cancer Institute is described by the NCIT as 'Subcommittee of the Board of Scientific Counselors, NCI'. At the same time, the NCIT lists the 'Board of Scientific Counselors' as a super-item of the subcommittee. The formal relation appropriately describing the relation between a sub-item and its super-item (e.g. a species and its genus) is the 'is a' relation. But, according to the NCIT-definition, 'Subcommittee B Basic Sciences *is_a* Board of Scientific Counselors, NCI' is not a true proposition. It is indeed false or even nonsense because both terms are proper names of individual institutions. In order to be meaningful, the *is_a*-relation requires general terms on both sides, i.e. terms that allow for multiple instantiations. Thus, already for syntactic reasons, the Subcommittee cannot be a sub-type of the NCI Board of Scientific Counselors. Nor can the Subcommittee be an instance of the Board, for the Board is a particular and not a universal (and only universals can have instances). Presumably, the intended meaning is that all Subcommittee members are Board members or that Subcommittee membership requires Board membership, but this cannot be represented by a subsumption relation between the respective institutions.

As can be drawn from Table 16.2, over and above universals or types, the NCIT does also contain terms for particular institutions, like the American Cancer Society and the NCI itself. The NCIT is even self-referential in so far as it lists itself among other particular information resources like the Gene Ontology or the medical online bibliography PubMed. Normally, particulars do not feature within an ontology, which is sometimes characterized as a hierarchy of universals only. Nevertheless, individuals like those that are mentioned within the NCIT are systematically related to these universals and are thus important for ontology. Such systematic relations between entities from Table 16.2 (and others) can be represented by means of formal ontological relations like those suggested by BFO or the OBO relational ontology.[37] In Table 16.3 I list some examples for assertions of basic ontological relations using terms from the NCIT for the relata.

Apart from the *is_a*-relation, these formal ontological relations are missing from the NCIT. The part-of-relation for continuants is only defined for physical or conceptual parts and even the *is_a*-relation is quite often used very strangely, as our discussion in the preceding sections has shown. Rigorous application of both a coherent set of top-level categories and such ontological relations will greatly improve the representation of social entities in general and in the NCIT in particular.

Table 16.3: Ontological relations and social entities

Individuals are instances of universals	University of California at Santa Cruz **instance_of** *University* NCI Thesaurus **instance_of** *Thesaurus* Health Insurance Portability and Accountability Act **instance_of** *Law*
Universals are subsumed by higher-order universals	*College is_a School* *Doctorate Degree is_a Academic Degree* *Application is_a Document*
Universals are (not necessarily directly) subsumed under their top-level categories	*University is_a Independent Continuant* *Doctorate Degree is_a Dependent Continuant* *Conference is_a Occurrent*
Dependent continuants inhere in independent continuants	*Employment inheres_in Employee* *Culture inheres_in Population Group* *Marital Status inheres_in Person*
Occurrents have independent continuants as agents and participants	*Research project has_agent Research unit* *Grant application has_agent Scientist* *Exchange Programme has_participant Scientist*
Independent continuants can be part of other independent continuants	*Research unit has_part ScientistUniversity has_part Department* *Document_body part_of Document*

Conclusion

The aim of this essay was to show how social ontology can help to classify social entities in medical information sciences. Using the current research in social ontology as background, I presented four rules for the classification of social entities and applied them to examples from the NCIT. Using BFO and OBO standards, I then developed a small fragment of an ontology of social entities to demonstrate how, together with these standards, the rules suggested here can help to improve the representation of social entities in information systems.

Acknowledgements

Part of the research for this paper was done under the auspices of the Wolfgang Paul Program of the Alexander von Humboldt Foundation and has been supported by the Volkswagen Foundation under the auspices of the project 'Forms of Life'. Thanks to Barry Smith, Niels Grewe and several anonymous referees for comments on earlier versions of this essay.

17 INFO-COMPUTATIONALISM AND PHILOSOPHICAL ASPECTS OF RESEARCH IN INFORMATION SCIENCES

Gordana Dodig-Crnkovic

Introduction

Historical development has led to the decay of natural philosophy, which until the nineteenth century included all of our knowledge about the physical world, in the growing multitude of specialized sciences, within the 'Classical Model of Science'. The focus on the in-depth enquiry disentangled from its broad context led to the problem of the loss of a common world view and the impossibility of communication between specialist research fields because of different languages that they developed in isolation. The need for a new unifying framework is becoming increasingly apparent, with information technology enabling and intensifying the communication between different research fields, knowledge communities and information sources. This time, not only natural sciences, but also all of human knowledge is being integrated through a global network such as the internet with its diverse knowledge and language communities.

Info-computationalism (ICON) as a synthesis of pancomputationalism and paninformationalism presents a unifying framework for the understanding of natural phenomena including living beings and their cognition, their ways of processing information and producing knowledge. Within ICON, the physical universe is understood as a network of computational processes on an informational structure. The matter/energy in this model is replaced by information/computation where information is the structure, whose dynamics are identified as natural computation.

ICON is an example of a philosophical framework in a direct connection with the related scientific fields, and the process is one of mutual exchange: scientific findings influence philosophical thinking, and vice versa. Research is ongoing in natural computing on modelling natural phenomena including living organisms as info-computational agents and implementing natural computation principles on technological artefacts. Lessons learned from the design and implementation of our understanding of living natural computational agents leads to artefacts being increasingly capable of simulating the essential abilities of living organisms to process and structure information. Among other things, ICON supports scientific understanding of the mind (perception, thinking, reasoning, will, feelings, memory, etc.) providing computational naturalist models of cognition.

Science and Philosophy

> Our best efforts are directed at finding out why things are as they are or why the events around us occur as they do. This is so in all disciplines and philosophy is no exception in this regard; what sets philosophy apart is that it probes deeper as well as being more general. It queries the presuppositions other disciplines leave untouched, and in trying to clarify such presuppositions and trace their inter-connections it seeks to find out how the world is put together and how it works.[1]

Knowledge, both propositional and non-propositional, is the basic constituent of science produced by the research process. Nevertheless, often knowledge is considered as identical with propositional knowledge and science is identified with a search for *truth* about the world; *truth* considered to be in a propositional form. History, sociology and philosophy of science all offer good reasons for seeing science as a goal-driven human activity aimed at production of *models* which enable us to predict, correlate and structure (compress, according to Chaitin) relevant information about the world.

The role of science as an information compression mechanism (in the sense of Chaitin's algorithmic information[2]) is especially visible in its search for simple and universal laws, which especially characterize physical sciences. Historically, however, the idea of natural law has evolved significantly. The ancient idea of *deterministic order* of the universe was closely related to the *principle of causality* that denotes a necessary relationship between one event (cause) and another event (effect) which is the direct consequence (result) of the first. However, *indeterminism* of quantum physics induced new elements into the picture of natural laws. Later on, *disorder* was found even in rule-governed systems, showing that deterministic functions can generate unpredictable results.

What is crucial to scientific knowledge is not its *certainty*, otherwise not much would qualify as knowledge in the history of science. Even the understanding of fundamental scientific ideas such as time, space, mass and trajectory have been successively historically revised. What has constantly been characterizing sciences is the *rationality* of their approaches, presuppositions and aims. Science is primarily an explanatory and predictive *tool of making sense*.[3]

Nowadays, the complexity of real-world problems has become a focus of the sciences, in the first place thanks to the computational capabilities of ICT. Instead of static, symmetric and steady-state, eternal order, a new dynamic picture emerges in which everything changes, and order is created as a pattern over layers of underlying physical processes, and very simple basic rules can lead to the evolution of complex systems.

The 'Classical Model of Science' and the Complexity of the Real World

When analysing the relationship between philosophy and science it is important to recognize the paradigm shift going on in both sciences and philosophy as a consequence of the recent growth of trans-disciplinary, interdisciplinary and cross-disciplinary research unfamiliar with classical philosophical analysis. Research practices and the resulting sciences today are not what they used to be before the ICT revolution. The effective information processing, storage and exchange capabilities of today's networked global research communities enable complex problems to be addressed, something that was impracticable previously because of the problems' informational and computational complexity. Such problems are typically defined at several levels of abstraction (levels of description, levels of functionality) and thus they usually cover several classical research disciplines.

This development towards complex problems made some presuppositions of existing sciences explicit. Among others, their domains of validity have become visible, making distinctions and relationships between different fields clearer and easier to recognize. A typical example where the domain dependence of different sciences becomes apparent is in the modelling of living organisms.

At the basis of our common intuitive understanding there is an idealized view of sciences, the 'Classical Model of Science', which according to Betti and de Jong[4] is a system S of propositions and concepts satisfying the following conditions:

(1) All propositions and all concepts (or terms) of S concern a specific set of objects or are about a certain domain of being(s).

(2a) There are in S a number of so-called *fundamental concepts* (or terms).

(2b) All other concepts (or terms) occurring in S are *composed of* (or are *definable from*) these fundamental concepts (or terms).

(3a) There are in S a number of so-called *fundamental propositions*.

(3b) All other propositions of S *follow from* or *are grounded in* (or are *provable* or *demonstrable from*) these fundamental propositions.

(4) All propositions of S are *true*.

(5) All propositions of S are *universal* and *necessary* in some sense or another.

(6) All propositions of S are *known to be true*. A non-fundamental proposition is known to be true through its *proof* in S.

(7) All concepts or terms of *S* are *adequately known*. A non-fundamental concept is
     adequately known through its composition (or definition).

The Classical Model of Science is a reconstruction a posteriori and sums up the
historical philosopher's ideal of scientific explanation. The fundamental is that
'All propositions and all concepts (or terms) of S concern *a specific set of objects or
are about a certain domain of being(s).*[5]

   This view of science, together with exponential growth of scientific knowledge, has led to the extreme compartmentalization and specialization of sciences
which can work well in some cases, while in others its aim and meaning may be
questioned. One example might be medicine with its many narrow specializa-
tions where a patient is treated by different specialists who see her/his health in
only one specific domain and administer medicines as if the patient is a sum of
disjointed domains and not one single organism in whom all different domains
overlap and interact. This compartmentalization has its historical roots in the
limitations of human information-processing capabilities. Today's info-computational networks present good tools for information processing and exchange
and they enable the building of complex knowledge structures in which experts
with different specialties can adaptively interact and make well-informed decisions, taking into account existing knowledge from other fields.

   ICT is already affecting the way research is done and scientific knowledge is
produced which will make the 'Classical Model of Science' just an element of a
complex knowledge production structure.

   After the age of idealizations and compartmentalization, sciences are starting
to examine their own roots, presuppositions and mutual relationships (contexts
awareness). Computer-aided science resembles electrified industry – a new
world of possibilities presents itself compared to the pre-ICT world. Nowadays,
information and knowledge management technology sciences can afford to take
into account and model the complexity of the real world with a huge variety of
parameters, complex structures and intricate dynamics. Instead of taking idealized frozen slices of reality, we can make realistic models, simulate and study
the dynamics of complex systems and their interactions. The step from the 'old'
'pre-computational' world to the new ICT world is not a trivial one. Often conceptual confusions arise about different kinds of knowledge, the domain of its
applicability, the underlying presuppositions, the relationships with other possible knowledge and similar.

Info-Computationalism as a Framework for Unity of Knowledge

Info-computationalism is a view according to which the physical universe on a
fundamental level can be understood as an informational structure, the dynamics of which is a computational process. The matter/energy in this model is

replaced by information/computation; matter (structure) corresponds to information while the dynamics – constant changes in the informational structure – are computational processes. In this view the universe is a huge computer network which by physical laws 'computes' its own next state (see Chaitin on pancomputationalism). Information is the fabric of the universe. An instantaneous 'snap' of the universe reveals the structure. Changes are computational processes. Computation is simply information processing.

ICON unites pancomputationalism (natural computationalism) with paninformationalism. In short, info-computationalism is a dual-aspect approach based on two fundamental concepts: information and computation. In this view, as many computationalists have already declared (among them Zuse, Fredkin, Wolfram, Chaitin and Lloyd), the universe is a huge computing system (or a network of computing processes) which computes its own next state by implementing physical laws. *It must be pointed out that the computing universe is not identical with (or reducible to) today's computers.* Rather, computing is what the whole of the universe does while processing its own information by simply following natural laws.[6] One might suspect that the computationalist idea is vacuous, and if everything is info-computational, then it says nothing about the world. The computationalist claim, however, should be understood as being similar to the claim that the universe is made of atoms. 'Atom' is a very useful concept which helps us to understand the world in many fields. So, too, is the info-computational approach. The universe is NOT 'nothing but atoms', but on some view (level of organization, level of abstraction) it may be seen as atoms.

Unlike many other (pan)computationalists, I do not presuppose that computationalism necessarily implies digital computing. As Seth Lloyd points out, on the basic quantum-mechanical level discrete and analogue, digital and continuous computing is going on.[7]

As already emphasized, physical reality can be addressed at many different levels of organization. Life and intelligence are the phenomena especially characterized by info-computational structures and processes. Living systems have the ability to act autonomously and store information, retrieve information (remember), anticipate future behaviour in the environment with help of information stored (learn), adapt to the environment (in order to survive). In 'Epistemology Naturalized'[8] I present a model which helps us to understand mechanisms of information processing and knowledge generation in an organism. Thinking of us and the universe as a network of computational devices/processes allows us more easily to approach the question about boundaries between living and non-living beings.

Info-computationalism sees our bodies as advanced computational machines in constant interaction with the environmental computational processes. Our brains are informational architectures undergoing computational processes on many levels of organization. On the levels of basic physical laws there is a computation going on. Everything that physics can conceptualize, describe, calculate, simulate and predict can be expressed in info-computationalist terms. On the level of molecules (with atoms and elementary particles as structural elements) there are computational processes going on. The nerve-cell level can be understood as the next level of relevance in our understanding of the computational nature of the brain processes. Neurons are organized in networks, and with neurons as building blocks, new computational phenomena appear on the level of the neural network. The intricate architecture of informational structures in the brain implements different levels of control mechanisms not unlike virtual machines on higher level running on the structure below. What we call 'informational architecture' is fluid and interactive, not so much a crystal-lattice-type rigid construction but more like networks of agents.

The development is occurring in two directions: analysing living organisms as info-computational systems/agents, and implementing natural computation strategies (organic computing, biocomputing) into artefacts. Lessons learned from the design and implementation of our understanding of living natural computational agents through iterative process of improvements will lead to artefacts that to an increasingly higher degree will be capable of simulating characteristics of living organisms, such as cognition.

Information and Computing Sciences and Philosophy

When discussing the relationship between philosophy and sciences, an instructive example can be found in the computing and philosophy (CAP) field. The following is the list over some of research fields presented at CAP conferences: *philosophy of information, philosophy of computation, computational approaches to the problem of mind, philosophy of computing, real and virtual modelling, simulations, emulations, computing and information ethics, societal aspects of computing and IT, philosophy of complexity, computational metaphysics, computational epistemology, computer-based learning and teaching.* From the list it is evident that CAP represents a forum for cross-disciplinary, interdisciplinary and multi-disciplinary knowledge exchange and the establishment of relationships between existing knowledge fields, and philosophical reflection over them. This development of a new body of knowledge is related to a distinct paradigm shift in the knowledge production mechanisms.[9] Globalization, information networking, pluralism and diversity expressed in the cross-disciplinary research in a complex web of worldwide knowledge generation are phenomena that need to be addressed on a high level of abstraction, which is offered by philosophical discourse. Examples of philosophical approaches closely connected to the ongoing paradigm shift may be found in contemporary works.[10]

In order to understand various important facets of ongoing info-computational turn and to be able to develop knowledge and technologies, dialogue and research on different aspects of computational and informational phenomena are central. Taking information as a fundamental structure and computation as information processing (information dynamics) one can see the two as complementary, mutually defining phenomena. No information is possible without computation (information dynamics), and no computation is possible without information.[11]

Knowledge as a Complex Informational Architecture: The Necessity of a Multi-disciplinary Communication

Why is it important to develop a multi-disciplinary discourse which will present the departure from the monolithic 'Classical Model of Science' caused by diversity of the domains, methods and levels of organization/levels of abstraction? The main reason is epistemological – multi-disciplinarity provides the fundamental framework suitable for common understanding and communication between presently disparate fields. This argument builds on a view of knowledge as informational construction. According to Stonier,[12] data is a series of disconnected facts and observations, which is converted into information by analysing, cross-referencing, selecting, sorting and summarizing the data. Patterns of information, in turn, can be worked up into knowledge which consists of an organized body of information. Stonier's constructivist view emphasizes two important facts:

- going from data to information to knowledge involves, at each step, an input of work, and
- at each step, this input of work leads to an increase in organization, thereby producing a hierarchy of organization.

Research into complex phenomena[13] has led to an insight that research problems have many different facets which may be approached differently at different levels of abstraction and that every knowledge field has a specific domain of validity.

This new understanding of a multidimensional many-layered knowledge space of phenomena have among others resulted in an vision of an ecumenical conclusion of science wars by recognition of the necessity of an inclusive and complex knowledge architecture which recognizes importance of a variety of approaches and types of knowledge.[14] Based on sources in philosophy, sociology, complexity theory, systems theory, cognitive science, evolutionary biology and fuzzy logic, Smith and Jenks present a new interdisciplinary perspective on the *self-organizing complex structures*. They analyse the relationship between the process of self-organization and its environment/ecology. Two central factors are the role of information in the building of complex structures and the development of topologies of possible outcome spaces. The authors argue for a continuous development from emergent complex orders in physical systems to cognitive capacities of living organisms, complex structures of human thought and to cultures. This is a new understanding of the unity of interdisciplinary knowledge, *unity in structured diversity*.[15]

In a complex informational architecture of knowledge, logic, mathematics, quantum mechanics, thermodynamics, chaos theory, cosmology, complexity, the origin of life, evolution, cognition, adaptive systems, intelligence, consciousness, societies of minds and their production of knowledge and other artefacts … there are two basic phenomena in common: *information and computation*, which provide a framework for those jigsaw puzzle pieces of knowledge to be put together into a complex and dynamic unified info-computational view.

The body of knowledge and practices in computing and information sciences, as a new research field, has grown around an artefact – a computer. Unlike old research disciplines, especially physics which has deep historical roots in natural philosophy, research tradition within the computing community up to now has been primarily focused on problem solving and has not developed very strong bonds with philosophy.[16] The discovery of the philosophical significance of computing in both philosophy and computing communities has led to a variety of new and interesting insights on both sides.

The view that information is the central idea of computing/informatics is both scientifically and sociologically indicative. Scientifically, it suggests a view of informatics as a generalization of information theory that is concerned not only with the transmission/communication of information but also with its transformation and interpretation. Sociologically, it suggests a parallel between the industrial revolution, which is concerned with the utilizing of energy, and the information revolution, which is concerned with the utilizing of information.[17]

The Relevance of Philosophy for Sciences and Sciences for Philosophy

The development of philosophy is sometimes understood in the way that it has defined new research fields and then left them to sciences for further investigations.[18] At the same time, philosophy traditionally also learns from sciences and technologies, using them as tools for production of the most reliable knowledge about the factual state of affairs of the world. We can mention a fresh example of current progress in modelling and simulation of brain and cognition that is of vital importance for the philosophy of mind. As has been the case so many times throughout history, the first approach that comes along when scarce empirical knowledge exists is the intuitive one that, however, does not need to be the best.

Wolpert, for example, points out that science is an *unnatural mode of thought*,[19] and it very often produces a counterintuitive knowledge, originating from the experiences with the world made by tools different from everyday ones, experiences in micro-cosmos, macro-cosmos and other areas hidden for our unaided cognition.

A good example of the 'unnatural' character of scientific knowledge is the totally counterintuitive finding of astronomy that the earth revolves around the sun and not the other way around, as our intuitions would tell us. At present, a similar Copernican revolution seem to be going on in the philosophy of mind, epistemology (understood in informational terms), in philosophy of information, and philosophy of computing. The recently published book, *Every Thing Must Go: Metaphysics Naturalized*,[20] rightly argues for philosophy (and specifically metaphysics) informed by the latest developments in special sciences, instead of philosophers' *a priori* intuitions and common sense. Specialist sciences such and cognitive science and neuroscience, for example, have collected valuable knowledge that should be adopted by philosophy of mind instead of relying on historical ideas based on common sense. The process going on in the opposite direction, from philosophy to specialist sciences, has been mentioned several times in this essay in examples of info-computationalism. Additionally, ethics must be mentioned as a branch of philosophy that is of increasing importance for sciences.

Sciences (*Wissenschaften*) are nowadays understood as rational tools for information compression which enable us to predict and efficiently handle different phenomena, and not (as previously believed) a source of absolute truth about the universe revealing the divine plan while reading the 'book of nature'. Instead, they are telling us at least as much about who we are in the relationship with the world – in that new light a conscious reflection about our values, motives, priorities and ways is vital. Especially when it comes to applied science and technology, the relevance of ethical analysis is obvious. Computers as our primary tools of information processing in research and otherwise can be used for good or bad and newly developed technologies can be misused. The challenge of radically new applied sciences and technologies is in what James Moor calls 'policy vacuums'[21] – it is not often possible to predict in what ways technology can be misused and what end results it might have. Ethical judgement is absolutely necessary and, in the long run, even in the education system, research ethics, computing ethics, information ethic and other specific branches of applied ethics accompanying specialized sciences and technologies should become an integral part of education and research system.

Answering Some of the Criticism Usually Made against Info-Computationalism

The info-computationalist view may be interpreted as a claim that the whole of the world is 'nothing but a (computational) machine' and that we humans are essentially machines with no free will or feelings. That is obviously not the case. The view that the universe is an info-computational phenomenon means that the universe as it is may be understood and modelled in info-computational language. Feelings, qualia and other mental phenomena are emergent properties of the physical world which is info-computational.

The role of different paradigms in our understanding of the universe can be analysed by means of historical examples. In the past, several major paradigm shifts occurred: from mytho-poetic to mechanistic to the emerging computationalist understanding of the universe. Consequently, we can ask the same question about the mytho-poetic and mechanist universe: was that understanding of the universe true? Was it real or merely metaphoric? For the mytho-poetic universe, the answer is simple – it was a metaphor. Even though mechanicism was primarily the view of inanimate matter, and mechanistic approaches to robotics (mechanical quasi-humans) did not work for any other purpose but entertainment, the mechanistic world view nevertheless helped us to learn a lot about the universe. It helped us to learn about inanimate things, but also about many basic facts of the living world (for example, that *there is no Élan vital* but the same physical mechanistic laws govern the whole of the physical universe, living as well as non-living).

The parallel development in the course of computationalism is ongoing. We will learn about the informational and computational resources and capabilities of the universe and we will develop even more powerful ways of learning, via intelligent systems that we will successively improve.

Knowing that biological organisms (including humans) are information-processing 'machines' does not make them any less fascinating, in the same way that knowing that all of us are made of atoms does not mean that we do not have free will, imagination and real feelings. Understanding fundamental level processes does not make music, arts and philosophy obsolete.

Info-computationalism helps us both by supplying the tools for knowledge and artefact production and also by supplying tools for understanding natural phenomena and artefacts on many different levels. That is also why philosophy is coming back to sciences based on info-computational knowledge. A holistic, high level of abstraction view is necessary as a self-reflective process of knowledge.

In sum: info-computationalism is by no means the *final answer* to 'Life!' 'The Universe!' and 'Everything!' (which is forty-two, as we learned from Douglas Adams as quoted by Vincent Müller in this volume, see p. 213), but a learning tool which will help us to again reach a unity of knowledge on one specific level of abstraction: the info-computational level.

So the answer (info-computationalism) is not *the final answer*, but all the same, it seems to be a reasonable answer to the reasonable question of how to get a common language for disparate specialist fields which can enable mutual understanding and the building of new knowledge, especially about living organisms and their processes such as cognition, mind and intelligence.

18 PANCOMPUTATIONALISM: THEORY OR METAPHOR?

Vincent C. Müller

Prelude: Some Science Fiction on the Ultimate Answer and The Ultimate Question

Many many millions of years ago a race of hyperintelligent pan-dimensional beings (whose physical manifestation in their own pan-dimensional universe is not dissimilar to our own) got so fed up with the constant bickering about the meaning of life which used to interrupt their favourite pastime of Brockian Ultra Cricket (a curious game which involved suddenly hitting people for no readily apparent reason and then running away) that they decided to sit down and solve their problems once and for all.

And to this end they built themselves a stupendous super computer ...

'O Deep Thought computer', Fook said, 'the task we have designed you to perform is this. We want you to tell us ... ' he paused, 'the Answer!'

'The Answer?' said Deep Thought. 'The Answer to what?'

'Life!' urged Fook. 'The Universe!' said Lunkwill.

'Everything!' they said in chorus.

(At this point the whole procedure is interrupted by two representatives of the 'Amalgamated Union of Philosophers, Sages, Luminaries and Other Thinking Persons' who demand to switch off the machine because it endangers their jobs. They demand 'rigidly defined areas of doubt and uncertainty!', and threaten: 'You'll have a national Philosopher's strike on your hands!' This is resolved by Deep Thought who says it will take 7.5 million years to resolve the question and observes that, in the meantime 'So long as you can keep disagreeing with each other violently enough and maligning each other in the popular press, and so long as you have clever agents, you can keep yourself on the gravy train for life'. This convinces the philosophers and they leave. – 7.5 million years later, Deep Thought answers:)

'All right', said Deep Thought. 'The Answer to the Great Question ... '

...

'Is ... ' said Deep Thought, and paused.

...

'Forty-two', said Deep Thought, with infinite majesty and calm ...

'Forty-two!' yelled Loonquawl. 'Is that all you've got to show for seven and a half million years' work?'

'I checked it very thoroughly', said the computer, 'and that quite definitely is the answer. I think the problem, to be quite honest with you, is that you've never actually known what the question is'.[1]

What is the Question? A Starting Point

Pancomputationalists say that the first story is literally true, in fact only part of the truth: not only is the earth a computer, so is everything else; the universe is a computer.[2] In the following, I will try to learn a lesson from this strange story. Rather than investigate the truth of this answer, I will try to understand what it might mean – which will force me to speculate about what the question really is (a much harder problem than finding the answer, as we just learned).

A specification of what it means to say 'everything is a computer' is particularly urgent because this view is in acute danger of being devoid of any meaning. This is not just the usual analytic philosopher's question 'What do you mean?' – a very good question – but a particular danger for any theory that says 'everything is x'. If everything really is x, then the defender of the theory cannot point to some samples that are x and then to other samples that are non-x to explain the theory. So, the defender has to explain under what conditions, counterfactually, something would not be x. If this is not done, the good old Karl Popper test of being in principle falsifiable is not passed and the theory is in acute danger of being devoid of any meaning. My impression is that the need for this explanation has been overlooked, in the enthusiasm about the explanatory power of the new all-encompassing theory.

The main question about pancomputationalism is thus what it might mean by saying that the universe is a computer. This is clearly not the only question, however; in particular I would expect there to be empirical questions to determine the truth of the theory.

It seems that there are two quite distinct traditions in pancomputationalism, namely a realist and an anti-realist one. Furthermore, the realist tradition involves theorists that use a very wide notion of computing, and others that understand the notion of computing in the traditional sense of *digital* computing, thus arguing for the stronger thesis that the world is ultimately digital.

Two Theories and a Metaphor

Imagine a ball bouncing up and down, finally coming to a rest. The pancomputationalist remarks: 'This is all computation!' This remark has (at least) three interpretations:

(a) At any given time, the future states of the ball can be usefully described as the computational result from its present state (given all relevant factors).

(b) At any given time, the future states of the ball can be explained as the computational result from its present state (given all relevant factors).

(c) At any given time, the future states of the ball are computed from its present state (given all relevant factors). Bouncing is nothing but computation.

What (a) says is just that under some meaning of 'computation' it may be useful to describe the process as computational – I call this 'metaphor' below.

What (b) says is that a deterministic physics is true and can be expressed mathematically – I call this 'Theory II' below.

What (c) says is that the bouncing of the ball can be reduced to computing – I call this stronger view 'Theory I' below.

Anti-Realist Pancomputationalism

A venerable tradition in the philosophy of computing answers the question 'which entities in the world compute?' with the remark 'it depends on how you describe it'. This view is sometimes called pancomputationalism because it says that anything in the world can be described as a computer, if we so please. Versions of this tradition are represented, for example, by David Chalmers,[3] John Searle[4] and Oron Shagrir.[5] As far as I can tell, this theory does not make any claims on the ultimate computational nature of the world or of the computational theory of the world; in fact it is often used to argue against a theory that the mind is computational in any substantial sense. In the following, I will deal with a stronger theory, what I call 'realist' pancomputationalism. (Perhaps my arguments are relevant for this weaker theory also, but I do not investigate this here.)

I happen to think that anti-realism is not the right stance in respect to the question of which systems in the world are computers, because I think that the underlying digital states can be individuated without invoking any observer or person with intentions, but this question is beside the current investigation.[6]

Note that 'Theory II' is not a version of anti-realist pancomputationalism because the latter does not make a claim about the explanatory power of a description as computer.

Realist Pancomputationalism

There is an increasingly common view that the notion of computation can and should be used to describe many if not all physical processes, in fact that the physical world is at bottom computational. Gordana Dodig-Crnkovic says that, quite simply, 'every natural process is computation in a computing universe'.[7] Gregory Chatin concurs: 'The entire universe ... is constantly computing its future state from its current state, it's constantly computing its own time-evolution! And ... actual computers like your PC just hitch a ride on this universal computation!'[8] Accordingly, the traditional foundational notions of matter and energy are supposed to be replaced by computation.

A new version of this theory is info-computationalism, namely the view that the physical universe *can best be understood* as *computational* processes operating on *informational* structure. Classical matter/energy in this model is replaced by information, while the dynamics are identified as computational processes. On the face of it, this thesis is stronger than mere pancomputationalism, since it is the conjunction of pancomputationalism with a claim about the substance on which computing takes place. There are also tendencies to express info-computationalism as an epistemic thesis, which would indicate that info-computationalism cannot be classified under the realist positions, after all. (See the chapter by Dodig-Crnkovic in this volume, pp. 201–11.)

Within this framework, everything is computationalist. What is going on at the basic levels of physics and conventionally conceptualized, described, calculated, simulated and predicted in physics can be expressed in info-computationalist terms. Our bodies are advanced computational machines, at various levels, in constant interaction with other 'environmental' computational processes. Human nerve cells interact with each other and form complicated networks, thus producing another level of computational processes. At a yet higher level, these processes can be said to be processing representations of the outside world and result in events that are conscious to the agent, e.g. in what is called 'thinking'.

The traditional picture of computer science in general and of *artificial intelligence* in particular has been that the intelligence of humans lies in these higher level cognitive or 'intellectual' facilities and that the aim of engineering is to reproduce these cognitive structures in a different computational hardware, e.g. in 'artificial cognitive systems'. It has become increasingly obvious in recent decades, however, that this approach has agents standing on their heads, instead of their feet. A successful natural intelligent agent is an organism with an evolutionary history that stands in a multitude of computational relations to its environment, including other agents. Many of these processes will take place in complex systems and involve multiple agents or swarm intelligence. Pancomputationalism applies to the entire body of the agent, at various levels – the cognitive level is only one of many. Reproducing the computational structure would, in this picture, necessarily reproduce the 'emergent' properties of the natural agent. The wide pancomputationalist view will allow us to understand and to engineer a much larger range of processes than is commonly included in computer science and artificial intelligence.

Which precise formulation of the theory one adopts also depends on the notion of 'computation' that underlies these theories. Some theorists use a very broad notion, while some rely on a notion of digital computing that is identical to Turing's (or at least very closely related to it). If digital computing is taken as a base, this leads to the idea of a 'digital physics', based on an essentially discrete universe. For example: 'everything is made out of ⁰⁄₁ bits, everything is digital software, and God is a computer programmer, not a mathematician!'[9] This view goes back to John Wheeler's slogan 'It from Bit' and is developed in computer science by Edward Fredkin[10] and Stephen Wolfram.[11]

Pancomputationalism has repeatedly caught the public imagination, e.g. in *Wired* magazine[12] or at the recent very prominent Midwest KNS conference entitled 'What is Computation? (How) does Nature Compute?'

Computation

One of the many issues for pancomputationalism is to explain in which sense of computing the universe is a computer. We can, however, sidestep this issue for our argument. What we need is only the assumption that the pancomputationalist definition of computing, whatever it might be, will *include* classical Turing-machine computing. This kind of computing is a formal procedure that proceeds step-by-step and comes to a halt after a finite number of steps. The halting state is considered the 'output', and the procedure that is followed is an algorithm. Such procedures that reach a halting state after a finite number of formal steps are also often called 'effective procedure'.

Having said that, it clearly remains a desideratum for the pancomputationalist to clarify what she means by computing – particularly for the realist version of the theory, since the anti-realist version is already motivated by a particular view on computing.

A further desideratum is to explain in what sense some physical systems, like the machine on which I am writing this, are clearly computers and some processes inside them are clearly computational processes – while other things (the apple next to it) are not, and yet others may be, e.g. the human nervous system. Pancomputationalism as theory would not make these disputes trivial,[13] but it would have to explain how some systems can be computers on the basis of computers.

Pancomputationalism as a Theory I: The Universe is a Computer

Two Forms of Ontological Reductionism

Strong (Type) Reduction. A version of reduction that is not uncommon in the natural sciences is the discovery that one known property (e.g. temperature) is identical to another, basic, property (e.g. kinetic energy) – but that talk in terms of the basic property has advantages in terms of explanatory power. In this sense:

Property A is reducible to basic property B if and only if:
   Necessarily: If two objects are identical with B, they are identical with A, and inversely.

If this is the case, any (extensional) talk in terms of properties A can be replaced by talk in terms of properties B *salva veritate*. We might want to say, for example, that 'temperature is reducible to kinetic energy', in this sense. An 'identity theorist' in the philosophy of mind would say that 'having a mental state of type A is reducible to having a brain state of type B'.

As the discussion of identity theory shows, this kind of reductionism goes beyond showing that property B is somehow 'basic' in claiming that there can be only *one way* in which the basic property can produce the property A. So, in our case, only one type of computation can produce one type of physical property. This seems implausible and as far as I can tell, it is not defended by pancomputationalists. In any case, it will be sufficient to investigate a thesis that is implied by this strong reductionist thesis.

   Global supervenience. A promising explanation of the pancomputationalist reduction thesis is that the physical properties are based on computational properties, without requiring that a particular computational property will always be basic for the same physical property. After all, the hope is that a few computational properties will explain a lot of physics. This kind of relation can be specified with a notion that has become increasingly popular in the philosophy of mind: the notion of supervenience. Here it is supposed to capture the intuition that the mental is based on the physical, in the sense that two physically identical objects must share the same mental properties, but without requiring that there is only one physical way to bring about a particular mental property (this will allow for the same mental properties in beings that are physically very different). Since physics is concerned with law-like explanations, it seems apt to formulate our thesis as a general statement about *types* of properties, rather than as about particular *tokens*. Here is a classical formulation of supervenience, taken from Jaegwon Kim:

   For families of properties A and B: 'A *strongly supervenes* on B just in case, necessarily, for each x and each property F in A, if x has F, then there is a property in B such that x has G, and *necessarily* if any y has G, it has F'.[14] *Weak* supervenience is defined just the same, but without the second 'necessarily'.

Global supervenience for two sets of properties M and P, where M properties supervene on the more basic P properties:

'Any two worlds that are indiscernible with respect to their P properties are also indiscernible with respect to their M properties.'[15]

This means that if two objects are identical with basic property P, they must be identical with M, *but not inversely.* In other words, there can be no difference in M without a difference in P. Or else: M is *multiply realizable* with different Ps. Classical examples for this kind of supervenience are: 'Mental properties supervene on physical ones' and 'The look of a picture supervenes on its physical properties' (but the value does not, a physical duplicate of a painting by Rembrandt would look the same but have the value of a reproduction, not of a Rembrandt).

Reductive Info-Computationalism Produces Monsters

Let us now investigate the thesis that the physical processes are computational processes by saying that: 'Physical processes supervene on computation'. This means that if two physical processes P^1 and P^2 perform the same computation C^1, they are the same physical process.

This, however, seems to have absurd consequences. For one thing, it would mean that reproducing a computational process is to reproduce the physical process: reproducing the hurricane in the computational model produces a hurricane!

Second, for all we know about computation, it is not true that there is no difference in M (derived) without a difference in P (basic). Computing is multiply realizable: P^1 and P^2 can be two *different* physical processes but both compute *the same* C^1. This contradicts supervenience of P on C – and it indicates that pancomputationalists may have supervenience upside down: if anything, computation supervenes on physics (but see below).

I think it is useful to put this point in terms of 'levels of description' for a computer.[16] A computer can be described on at least three levels:

physical
syntactic
semantic

What we call computation takes place on the level of syntax, it is a purely formal procedure taking place in a physical mechanism, and perhaps having meaningful symbols. Put in terms of these levels, what Searle stressed is that syntax does not determine semantics (semantics does not supervene on syntax).[17] What I said above is that syntax does not determine physics (physics does not supervene on syntax).

David Deutsch, who uses the slogan 'The world is made of qubits' as a version of Wheeler's 'It from Bit' puts this point of multiple realizability thus:

> Universality means that computations, and the laws of computation, are independent
> of the underlying hardware. And therefore, the quantum theory of computation can-
> not explain hardware. It cannot, by itself, explain why some things are technologically
> possible and others are not.[18]

Computation is not constrained enough to explain physical reality, what he calls 'the hardware'.

Incidentally, it is not true either that physics determines syntax, in the sense that computation would supervene on physics: C^1 and C^2 can be two different processes but both be computed by the same P^1. This can be a matter of function or interpretation (this insight got anti-realist pancomputationalism off the ground).

Pancomputationalism as a Theory II: A Complete Theory of the Universe Can Be Formulated in Computational Terms

Given the failure of taking pancomputationalism as a theory of the universe, it should be said that some formulations suggest it is more of a theory of an expla-nation of the universe – perhaps this is what the theory really says, rather than just being a consequence of the reductive theory? I will take a quick look at this possibility, though it is somewhat speculative.

Presumably what this theory might say is that the formal process of comput-ing is sufficient for explanatory purposes, so it must claim that 'any process is formally describable'.

This would be good news for computer science, since it means that, in principle, anything can be programmed perfectly. Unlike in reductivist pancom-putationalism this programming of some natural process is only a simulation – but there are no practical limits to what can be achieved in this way.

Whether this view is true is surely a deep question for the philosophy of math-ematics, so allow me just to indicate why this position faces considerable obstacles.

Whatever precisely a formal description *is*, there are normally several possible formal descriptions for any given object or process. It is not clear that a particular one of these must be the right one (an impressive discussion is Putnam).[19] In particu-lar, a formal description would have to specify 'what is the point' of the description, but normal physical processes do not have functions by themselves (with the excep-tion of intentional processes and the possible exception of evolutionary processes), so the 'point' of the formalization must be provided by the observer.

For a computational description, this problem is probably even more serious, since even under a specific formal description of a process there are still several computations that it performs. If the description is '000110111001' and we are told in addition that the first 8 bits are 4 pairs of input and the last 4 are the output, then we might surmise that the process is addition. But this is underde-termined by the formal structure.

Pancomputationalism as Metaphor

What remains of pancomputationalism is its use as a metaphor, expressed in remarks like: 'The universe can often usefully be described as computational', 'The universe can often usefully be described as computing over information (as infosphere)' or 'Some of our scientific knowledge can be described in computational terms'. This metaphorical use will be a success if it is carefully distinct from more substantial philosophical theses. If this is ensured, it is very likely that many systems can usefully be described as computational, especially once a semi-formal description of their relevant factors has been achieved. This metaphorical use is insightful and important – we should just not stretch it into a theory.

19 THE IMPORTANCE OF THE SOURCES OF PROFESSIONAL OBLIGATIONS

Francis C. Dane

The study of philosophy provides many general benefits to members of any field or discipline, the easiest of which to defend are an appreciation of, and experience with, critical thinking, including the ability to apply principles thoughtfully and logically in a variety of contexts; it is the discipline that, according to Plato, Socrates believed made life worth living. Today, however, most disciplines can lay claim to critical thinking – information science certainly involves a great deal of logical analysis – but only philosophy, in the Western world, can lay claim to having developed logic and critical thinking[1] and thereby may have furthered the process more than any other discipline. Historically, philosophy is also the discipline in which one learns how to think about the most complex and important questions[2] including questions about what is right and proper; that is, philosophy arguably lays claim to the development of ethics.

Before going further, I should note that I am neither a philosopher nor an information scientist. I am a social psychologist and statistician whose interests have brought me into the realm of practical ethics primarily through ethical issues relevant to empirical research.[3] I should also note that I am firmly in the camp of those who consider there to be an important distinction between morals and ethics;[4] as do others, I argue that moral judgements essentially involve questions about whether or not rules, defined broadly, are followed, whereas ethical judgements essentially involve questions about whether or not a particular rule is worthwhile and, when there are incompatible rules, which rule should be granted higher priority.[5]

The distinction between morals and ethics is, I believe, more than mere semantics and is particularly important in the realm of practical ethics. It represents the difference between behaviour and theory, between concrete and abstract, between pronouncements and principles, between what one does and why one does it. Consider, for example, the experience of David Parnas, currently Professor of Computer Science and Information Systems at University of Limerick, who in the 1980s was asked to consult for the Panel on Computing in Support of Battle Management within the Strategic Defense Initiative Organization (SDIO) in the United States. Within a month of accepting the appointment, Parnas concluded that the task to which he was assigned was not possible. His colleagues agreed, but they also argued that the attendant funding would be useful for advancing information science and that those managing SDIO were willing to continue the project despite knowledge of his analysis. Therefore, they argued, the panel members were following the rules by fulfilling the contract; essentially, Parnas's colleagues argued they were fulfilling an informed client's wishes. Instead, Parnas resigned because he considered a principle, accepting funds under false pretences is unethical, to be more important than following rules, in the form of an ill-conceived contract or in the form of his colleagues' assertion that government funding should be accepted whenever it is offered. Instead of merely following the rules, Parnas based his behaviour on a principle and decided that the rules should not be followed.

As with critical thinking, no discipline has a monopoly on the study of ethics, as is evident in the plethora of professional organizations that have adopted codes of ethics.[6] Many, including those who have proposed specific codes, have argued, however, that codes are insufficient. Stephen Unger, for example, argued that codes should be used 'as a guide ... to suggest factors to be considered and to raise questions'.[7] John Ladd more strongly argued that 'codifying ethics ... makes no more sense than ... codifying medicine, anthropology or architecture'.[8] Nevertheless, professional and organizational codes of ethics are necessary exactly because they can be used to guide behaviour, because they can be used to present questions of concern, and because they can be used to force professionals to confront the ethical challenges inherent in their disciplines. While necessary, codes of ethics are not sufficient because they primarily contain rules, and those rules often can be incompatible. Even when codes contain principles, the principles are typically codified without context, without the logical arguments through which others have made those principles worthy of consideration. In short, codes of ethics do not contain ethics, per se; for that one must turn to philosophy and the ethical theories produced by members of that discipline.

Let us use the Software Engineering Code of Ethics and Professional Practice,[9] developed jointly by the Association for Computing Machinery (ACM) and Institute of Electrical and Electronics Engineers (IEEE), and hereafter referred to as 'the Code', as a running example to illustrate the importance of understanding how philosophy serves as the foundation for all codes of ethics. The Code contains eight aspirations or principles that 'identify the ethically responsible relationships in which individuals, groups, and organizations participate and the primary obligations within these relationships'. Each principle is illustrated with multiple descriptions of behaviours that are consistent with the principle. The developers recognize that the 'Code is not a simple ethical algorithm that generates ethical decisions', nor does it relieve software engineers of the requirement 'to use ethical judgement to act in a manner which is most consistent with the spirit of the Code'.[10] That is, the developers recognized that the Code, per se, was not sufficient, that additional training in ethics would be required in order to meet the aspirations within the Code. Whence this additional training? How does one understand the 'spirit' of the Code? The training and understanding must come from philosophy, of course, specifically from the theories proposed within philosophy to explain one's ethical obligations.

As we examine the Code, I shall offer specific ethical theories that can be used to provide a foundation for each aspiration. These theories are chosen, however, to illustrate what may be the most obvious theory relevant to the aspiration and not for the purpose of arguing that the chosen theory is the only, or perhaps even the best, theory to explain the ethical foundation of the aspiration. That is, my purpose is not to argue which philosophical theory of ethics is the most logical or the most useful theory; my purpose is to illustrate that one cannot understand any of the aspirations fully without reference to and knowledge of one or more ethical theories.

Act Consistently with the Public Interest

The first principle of the Code, promoting the public interest, is certainly not unique; indeed, Deborah Johnson, among many others, has argued that the public interest should always be the primary obligation of any professional organization because the public grants specific privileges and powers to professionals in exchange for the professionals' accomplishing necessary, esoteric social functions.[11] This, then, begs the question of what, exactly, the public interest might be. Much earlier, in his treatise on utility as the foundation of ethical considerations, Jeremy Bentham[12] made what was then the not-so-obvious and somewhat controversial point that 'the interest of the community ... is ... the sum of the interests of the several members who compose it'. Which community? Which members? Bentham again informed us that the community to be addressed must include all who are affected by the action or rule under consideration.

What Parnas understood, but his colleagues apparently did not, was that their community included more than each of them, more than the project managers who served as the client,[13] more than the United States, and more than the discipline of information science. Parnas understood that accepting funds for a task one cannot complete affects all other professionals negatively and therefore must be avoided even if a few may temporarily benefit.

Without sufficient study of the philosophy espoused by Bentham and those who followed in his philosophical footsteps, such as William James and John Stuart Mill, it is difficult to understand whose interests should be included in 'the public interest' and equally difficult to understand how much to consider the interests of each of the community's members. Does one, for example, consider the interests of those with whom one has a direct relationship to be more important than the interests of those with whom one has only a whisper of an indirect relationship? The answer to this straightforward question is beyond the scope of this essay, but is certainly attainable through a study of utilitarian ethics.

Similarly, pursuing this first principle of the Code requires one to determine the criteria by which one decides what is in the public interest. Bentham's well-argued criterion, 'the tendency it has to augment the happiness of the community',[14] belies the complexity associated with determining how to measure happiness. Does one, for example, consider the various causes of happiness to be of equal weight? Does one consider happiness from a parsimonious or a liberal viewpoint, from an egoistic or an empathetic viewpoint? Again, the answers to these and similar questions are beyond the purview of this essay, but certainly can be found through a consideration of the philosophy underlying utilitarianism and other theories of ethics.

Ensure Products and Modifications Meet the Highest Professional Standards

The third aspiration, to meet the highest professional standards, is common to many codes and mission statements[15] and commits one to do more than a good-enough job; one is committed to strive for excellence, to be virtuous. The mere mention of 'excellence' or 'virtue' to anyone familiar with ethics brings to mind Aristotelian ethics, which has since become known as virtue ethics.[16] According to Aristotle, 'human good turns out to be activity of soul in accordance with virtue',[17] that is, in accordance with attaining excellence. From the Aristotelian perspective, one cannot become fully human, cannot achieve *eudaemonia*, without the development of character through habituation of virtuous actions. To become virtuous, one must make a rational decision to act virtuously, as opposed to justifying one's actions merely by claiming to be a virtuous person. Thus, it is not sufficient to know professional standards or to rest upon laurels achieved from having met them in the past; one is to aspire to meet the highest professional standards at all times in order to develop the habit of excellence.

But one also hears about Aristotle's concept of the Golden Mean, 'virtue must have the quality of aiming at the intermediate',[18] from which others argue (incorrectly) that excessive virtue, such as consistently striving for excellence, is to be avoided and that compromise is the preferred mechanism.[19] The tendency to mistake Aristotle's notion of 'an intermediate between defect and excess'[20] with the notion of compromise, the tendency to think that 'good enough' because one lacks time or motivation is the same as striving for excellence, is relatively common among individuals who have not given careful attention to Aristotle and others who have expanded on his theory.[21] Without such attention, one would have considerable difficulty grasping the underlying meaning of Aristotle's Golden Mean or, for that matter, of this third aspiration in the Code. Understanding why one should strive to meet the highest standards, and thereby have the ability to address this aspiration in all situations, requires knowledge of philosophy.

Integrity and Independence in Professional Judgement

The inclusion of integrity and independence in the fourth principle of the Code commits one to aspire to consistency of behaviour as well as to the maintenance of autonomy. Thus, one faces the difficult task of monitoring one's own behaviours so as to make them consistent with the values and principles one espouses while simultaneously preventing others from unduly influencing the principles on which those behaviours are based. This does not mean that one should adopt a dogmatic approach such that one ignores or carelessly denigrates those with whom one disagrees. One should, of course, consider the viewpoints of others, but one should do so critically, to ensure that one both understands the premises on which differing viewpoints are based and that one has analysed the logic underlying the development of those viewpoints from the premises.

The first part of the aspiration leads one toward a universality of action, toward the goal of a life without exceptions, but without a foray into philosophy one is not likely to understand the source of this aspiration. Those who have read philosophy will recognize in this aspiration Kant's[22] Categorical Imperative: 'Act only on that maxim whereby thou canst at the same time will that it should become a universal law'.[23] Thus, aspiring toward integrity obligates one to choose the values and principles that guide one's actions such that one would be willing to argue legitimately in favour of requiring everyone else, literally every human, to choose those same principles.

Because *post-hoc* rationalizations are unlikely to be consistent with existing principles, the aspiration of integrity commits one to develop *a priori* principles on which to base all of one's behaviour as well as to employ those principles rationally during any decision process pursuant to behaviour. We are reminded that Parnas based his decision regarding SDIO on principles he regarded as obligatory for every professional: responsibility, beneficence and realistic problem solving.

The independence portion of this aspiration leads us to the principle of autonomy, which to Kant was 'the basis of the dignity of human and of every rational nature'.[24] Autonomy involves an informed choice and one must therefore be able to reason objectively and logically in order to address this aspiration. Equally important is the need to obtain information with which to engage in rational decision-making. Independence, therefore, cannot involve ignoring others, nor can it involve uncritically accepting direction from others. Instead, independence must be based on accepting information from others and analysing that information critically in order to make an autonomous decision.

Much has been written about Kant's Categorical Imperative; a quick search of the online version of Philosopher's Index™ yielded more than five hundred articles, chapters and books. Similarly, a search for 'autonomy' yielded more than 4,600 entries. One would not expect a professional, except, perhaps, a professional ethicist, to have read all of that material, but it should be obvious that one must consult at least some of the philosophical literature if one expects to attain this aspiration of the Code.

Advance the Integrity and Reputation of the Profession

The fifth aspiration of the Code, the directive to advance the integrity and reputation of the profession, can easily be perceived as continuing the previous emphasis on integrity by expanding it to the entire profession. Thus, one should not only aspire to living a life without exceptions, one also should aspire to ensuring that one's colleagues do the same. In addition to Kant's Categorical Imperative, however, what comes to mind is the concept of moral responsibility, espoused by Graham Haydon as an addition to H. L. A. Hart's four classic types of responsibility (role, causal, liability, capacity). Of particular interest is John Ladd's concept of collective moral responsibility,[25] through which he argued that all of the members of a profession are non-exclusively responsible for the future of the profession; each member individually bears some responsibility for correcting the mistakes of the past and preventing mistakes in the future. One is called upon to aspire to avoid egoism,[26] even egoistic pursuit of integrity, in favour of pursuit of collective integrity, to accept some degree of responsibility for the actions of one's colleagues.

Ladd places collective moral responsibility clearly within virtue theory: to be responsible in this sense is a virtue that cannot be meaningfully predicated of ... a structure of rules, offices, jobs, etc.[27] Thus we see the relevance of Aristotelian ethics in a discussion of what may as well be considered an aspiration based on Kantian theory. Disentangling the admixture of Kantian and Aristotelian ethics obviously requires considerable study in philosophy. Positions concerning such an admixture range from a strong emphasis on Aristotelian theory[28] to an equally strong emphasis on Kantian theory[29] and various and sundry positions in between.[30] Even for so seemingly simple an aspiration as promoting integrity, additional reading is required.

Be Fair and Supportive

The combined emphasis on fairness and support in the seventh aspiration of the Code clearly challenges information science professionals to work from a spirit of cooperation extended to all within the profession. The examples provided in the Code for this aspiration include verbs such as encourage, assist and review, which presents no small challenge in a discipline that is well known for promoting advancement through competition.[31] How, then, should one cooperate in what is clearly a competitive discipline; how should one be fair?

The question brings to mind John Rawls's theory of justice and his focus on rights and freedoms as the basis of fairness.[32] First, being fair requires egalitarian structures; 'each person is to have an equal right to the most extensive scheme of equal basic liberties'.[33] Second, being fair requires attention to social inequalities; 'those better situated are just if and only if they work as part of a scheme which improves the expectations of the least advantaged members of society'.[34] Exactly why one should pursue egalitarian structures or ensure benefit for the least advantaged is, of course, well beyond our present scope, but careful consideration of Rawls's work, as well as the work of those who have commented upon and extended his work, such as Martha Nussbaum[35] and Amartya Sen, will provide answers.

Participate in Lifelong Learning and Promote an Ethical Approach to Practice

The eighth aspiration of lifelong learning and an ethical approach to practice, coupled with the fifth aspiration involving an ethical approach to management, could just as well have been written as 'study and use ethics'. The ACM/IEEE Joint Task Force clearly recognized that the Code was not sufficient, that additional exploration and study of ethics was necessary. Indeed, Clause 8.06 calls software engineers specifically to 'improve their knowledge of this Code, its interpretation, and its application to their work'.

One must ask, however, how one is to fulfill these aspirations. Whose ethics are to be applied? Which theories are to be studied? The preceding aspirations led us to theories of ethics from Aristotle, Bentham, Kant and Rawls, who cover centuries of theoretical development and each of whom have hundreds of commentators in favour of and in disagreement with their work. And, as I noted at the outset, these four are merely the first to come to mind for any given aspiration. Someone else may well have chosen four different theorists, or have argued that one theorist comprehensively addresses all of the aspirations.

At first glance, it may seem possible to address the meaning and implication of the principles within the Code without explicit reference to one or more theories of ethics. The behavioural examples provided in the code seem to make it possible that one could do so through an inductive process. However, when more than one of the aspirations direct one's attention to the study of ethics, it becomes clear that it is not possible to understand the Code as comprehensively as one could without studying the theories upon which the Code's foundation was built. Analogously, one could make certain *assumptions* about how a machine functions or how a programme works through analysis of its uses, but one cannot *know* how a machine or programme works without direct examination of the foundation, the component parts or the source code.

It should be clear, then, that the study of philosophy is essential to each person in the discipline of information science, at least each who aspires to serve the profession in an ethical manner. Indeed, I proffer that the study of ethics is essential to membership in any profession. If one is to accept the collective moral responsibility that attends the entry into and continued work within a profession, then one must accept the ancient call to live an examined life not only through lifelong study within one's profession but also through lifelong study of philosophy.

NOTES

Hagengruber and Riss, 'Introduction: Philosophy's Relevance in Computing and Information Science'

1. C. Allen and W. Wallach, *Moral Machines – Teaching Robots Right from Wrong* (Oxford: Oxford University Press, 2009); R. Capurro, 'Informationsethik. Eine Standortbestimmung', at http://www.capurro.de/infoethik_standort.htm [accessed 18 November 2013]; C. Ess and R. Hagengruber (eds), *The Computational Turn: Proceedings IACAP Aarhus* (Münster: MV Wissenschaft, 2011); L. Floridi (ed.), *The Cambridge Handbook of Information and Computer Ethics* (Cambridge: Cambridge University Press, 2011); C. Fuchs, *Internet and Society: Social Theory in the Information Age* (London: Routledge, 2008); C. Fuchs, K. Boersma, A. Albrechtslund and M. Sandoval (eds), *Internet and Surveillance: The Challenges of Web 2.0 and Social Media* (New York, London: Routledge, 2012); K. Mainzer, *Leben als Maschine? – Von der Systembiologie zur Robotik und Künstlichen Intelligenz* (Paderborn: Mentis Verlag, 2010); H. Hrachovec and A. Pichler (eds), *Philosophy of the Information Society: Proceedings of the 30th International Ludwig Wittgenstein Symposium* (Frankfurt: Ontos, 2008); P.-P. Verbeek, 'Technology and the Moral Subject', in P.-P. Verbeek, *Moralizing Technology: Understanding and Designing the Morality of Things* (Chicago, IL: University of Chicago Press, 2011), pp. 66–89; A. Beavers, 'Moral Machines and the Threat of Ethical Nihilism', in P. Lin, G. Bekey and K. Abney (eds), *Robotic Ethics: The Ethical and Social Implications of Robotics* (Cambridge, MA: MIT Press, 2011), pp. 333–4; C. Ess and M. Thorseth, 'Global Information and Computer Ethics', in Floridi (ed.), *The Cambridge Handbook of Information and Computer Ethics*, pp. 163–80.
2. R. Schaerer, *Episteme et techne: Etudes sur les notions de connaissance et d'art d'Homère à Platon* (Macon: Protat frères, imprimeurs, 1930); G. Vattimo, *Il concetto di fare in Aristotele* (Turin: Università di Torino, 1961); P. Pellegrin, 'Techne ed episteme', in S. Sethis (ed.), *I Greci. Storia cultura arte società* (Torino: Einaudi, 1997), pp. 1189–203; A. Balansard, *Techné dans les dialogues de Platon* (Sankt Augustin: Academia Verlag, 2001); J. Barnes, 'Aristotle's Theory of Demonstration', in J. Barnes, M. Schofield and R. Sorabji (eds), *Articles on Aristotle: 1* (London: Duckworth, 1975), pp. 65–87.
3. It was even followed by the conceptual separation of mind and body that influenced philosophical concepts as well as technical inventions.
4. Plato, *Meno*, cited according to *Plato in Twelve Volumes*, 12 vols, trans. W. R. M. Lamb (Cambridge, MA: Harvard University Press; 1967), vol. 3, pp. 97ff. See also http://

www.perseus.tufts.edu/hopper/text?doc=Perseus%3Atext%3A1999.01.0178%3Atext
%3DMeno%3Asection%3D97a. [accessed 6 November 2013].

5.   See H. Putnam, 'Mind and Machines', in S. Hook (ed.), *Dimensions of Mind* (New
     York: New York University Press, 1960) and H. Putnam, *Mind, Language, and Real-
     ity* (Cambridge: Cambridge University Press, 1975). See also A. Sloman, *The Computer
     Revolution in Philosophy: Philosophy, Science and Models of Mind* (Sussex: The Harvester
     Press, 1978).

6.   P. M. Churchland, *The Engine of Reason, the Seat of the Soul: A Philosophical Journey into
     the Brain* (Cambridge, MA: MIT Press, 1995).

7.   P. Langley, 'Rediscovering Physics with Bacon 3', *Proceedings of the Sixth International
     Joint Conference on Artificial Intelligence* (1979), pp. 505–7; P. Langley, H. A. Simon, G.
     L. Bradshaw and J. J. Zytkow, *Scientific Discovery: Computational Exploration of the Crea-
     tive Process* (Cambridge, MA: MIT Press, 1987); see also R. Hagengruber, 'Algorithmus
     und Kreativität', in G. Abel (ed.), *Kreativität. XX. Deutscher Kongress für Philosophie
     26.–30. September 2005 in Berlin* (Berlin: Universitätsverlag der TU Berlin, 2005), pp.
     235–7.

8.   K. Mainzer, 'The Emergence of Mind and Brain: An Evolutionary, Computational, and
     Philosophical Approach', in R. Banerjee and B. Chakrabati (eds), *Models of Brain and
     Mind: Physical, Computational, and Psychological Approaches* (Amsterdam: Elsevier,
     2007); see also C. Allen, 'Artificial Life, Artificial Agents, Virtual Realities: Technolo-
     gies of Autonomous Agents', in Floridi (ed.), *The Cambridge Handbook of Information
     and Computer Ethics*, pp. 219–23.

9.   W. Dilthey, *Einleitung in die Geisteswissenschaften* (Leipzig: Duncker & Humblodt,
     1883).

10. M. Heidegger, *Being and Time* (New York: Harper & Row, 1927); M. Heidegger, 'Die
     Frage nach der Technik', in M. Heidegger, *Reden und Aufsätze* (Pfullingen: Neske,
     1954), pp. 9–40; T. W. Adorno et al., *Der Positivismusstreit in der deutschen Soziologie*
     (Darmstadt: Luchterhand, 1978).

11. F. Dessauer, *Philosophie der Technik* (Bonn: F. Cohen, 1927); E. Cassirer, 'Form und
     Technik' , in E. Cassirer, *Symbol, Technik, Sprache. Aufsätze aus den Jahren 1927–1933*
     (Hamburg: Meiner, 1930), pp. 39–91.

12. See the famous arguments in R. Carnap, 'Die physikalische Sprache als Universalsprache
     der Wissenschaft', *Erkenntnis*, 2 (1932), pp. 432–65 and O. Neurath, 'Protokollsätze',
     in *Erkenntnis*, 3 (1932–3), pp. 204–14; R. Carnap, 'Über Protokollsätze', *Erkenntnis*, 3
     (1932–3), pp. 215–28; M. Schlick, 'Über das Fundament der Erkenntnis', *Erkenntnis*, 4
     (1934), pp. 79–99; G. W. Reichenbach, *Erfahrung und Prognose/Experience and Predic-
     tion* (Braunschweig: Vieweg, 1983).

13. W. v. O. Quine, *On What There is* (New York: Harper & Row, 1953); W. Stegmüller,
     Probleme und Resultate der Wissenschaftstheorie und Analytischen Philosophie, Band 2
     [Volume 2], *Theorie und Erfahrung* (Berlin and New York: Springer, 1974).

14. P. Feyerabend, *Against Method* (London and New York: New Left Books, 1975); T. S.
     Kuhn, 'Logic of Discovery or Psychology of Research', in I. Lakatos and A. Musgrave
     (eds), *Criticism and the Growth of Knowledge* (Cambridge: Cambridge University Press,
     1970), pp. 1–23.

15. T. Winograd and F. Flores, *Understanding Computers and Cognition: A New Foundation
     for Design* (Norwood, NJ: Ablex, 1986).

16. H. L. Dreyfus, *What Computers Can't Do: A Critique of Artificial Reason* (New York:
     Harper & Row, 1972).

17. R. Brooks, 'Achieving Artificial Intelligence through Building Robots', MIT Technical Report, 1986, at http://dspace.mit.edu/bitstream/handle/1721.1/6451/AIM-899.pdf [accessed 18 November 2013].

18. P. Thagard, *Computational Philosophy of Science* (Cambridge, MA: MIT Press, 1988).

19. P. Dourish, *Where the Action is: The Foundations of Embodied Interaction* (Cambridge, MA: MIT Press, 2001).

20. Y. Engeström, *Learning by Expanding: An Activity-Theoretical Approach to Developmental Research* (Helsinki: Orienta-Konsultit Oy, 1987).

21. L. S. Vygotsky, *Mind in Society: Development of Higher Psychological Processes* (Cambridge, MA: Harvard University Press, 1978); A. N. Leont'ev, *Problems of the Development of the Mind* (Moscow: Progress, 1981).

22. B. A. Nardi (ed.), *Context and Consciousness: Activity Theory and Human-Computer Interaction* (Cambridge, MA: MIT Press, 1995); V. Kaptelinin and B. A. Nardi, *Acting with Technology: Activity Theory and Interaction Design* (Cambridge, MA: MIT Press, 2006).

23. I. Nonaka and H. Takeuchi, *The Knowledge-Creating Company: How Japanese Companies Create the Dynamics of Innovation* (New York: Oxford University Press, 1995).

24. G. Ryle, *The Concept of Mind* (Chicago, IL: Chicago University Press, 1949); M. Polanyi, *Personal Knowledge: Towards a Post-Critical Philosophy* (Chicago, IL: University of Chicago Press, 1962); M. Polanyi, *The Tacit Dimension* (Garden City, NY: Doubleday, 1966).

25. T. Berners-Lee, J. Hendler and O. Lassila, 'The Semantic Web: A New Form of Web Content That is Meaningful to Computers Will Unleash a Revolution of New Possibilities', *Scientific American*, 284:5 (2001), pp. 34–43.

26. T. R. Gruber, 'A Translation Approach to Portable Ontology Specifications', *Knowledge Acquisition*, 5 (1993), pp. 199–220.

27. B. Smith, 'Beyond Concepts: Ontology as Reality Representation', in A. C. Varzi and L. Vieu (eds), *Formal Ontology in Information Systems: Proceedings of the Third International Conference (FOIS-2004), Frontiers in Artificial Intelligence and Applications* (Amsterdam: IOS Press, 2004), pp. 73–84; B. Smith, 'The Logic of Biological Classification and the Foundations of Biomedical Ontology', in D. Westerståhl (ed.), *Invited Papers from the 10th International Conference in Logic Methodology and Philosophy of Science* (London: King's College Publications, 2005), pp. 505–20; K. Munn and B. Smith (eds), *Applied Ontology: An Introduction* (Frankfurt: Ontos, 2009).

28. C. E. Shannon and W. Weaver, *The Mathematical Theory of Communication* (1948; Urbana, IL: University of Illinois Press, 2002).

29. L. Floridi, *Philosophy and Computing: An Introduction* (London and New York: Routledge, 1999); L. Floridi, 'What is the Philosophy of Information?', *Metaphilosophy*, 33:1–2 (2002), pp. 123–45; L. Floridi (ed.), *The Blackwell Guide to the Philosophy of Computing and Information* (Oxford: Blackwell, 2004); and P. Allo, *Putting Information First: Luciano Floridi and the Philosophy of Information* (Oxford: Blackwell, 2011).

30. L. Wittgenstein, *Tractatus logico-philosophicus* (London: Keagan Paul, 1922).

31. L. Wittgenstein, *Philosophical Investigations* (1953; Oxford: Blackwell, 1991).

32. L. Floridi, 'Semantic Conception of Information', in E. N. Zalta (ed.), *Stanford Encyclopedia of Philosophy*, 2005, at http://plato.stanford.edu/entries/information-semantic [accessed 14 September 2010].

33. R. Turner and A. Eden, 'The Philosophy of Computer Science', in E. N. Zalta (ed.), *Stanford Encyclopedia of Philosophy*, 2008, at http://plato.stanford.edu/entries/computer-science [accessed 21 November 2013].

34. D. Barker-Plummer, 'Turing Machines', in E. N. Zalta (ed.), *The Stanford Encyclopedia of Philosophy,* summer 2013, at http://plato.stanford.edu/archives/sum2013/entries/turing-machine [accessed 21 November 2013].

35. N. Immerman, 'Computability and Complexity', in Zalta (ed.), *Stanford Encyclopedia of Philosophy*, 2008, at http://plato.stanford.edu/entries/computability [accessed 29 May 2013].

36. S. Horst, 'The Computational Theory of Mind', in E. N. Zalta (ed.), *Stanford Encyclopedia of Philosophy*, 2009, at http://plato.stanford.edu/entries/computational-mind [accessed 29 May 2013].

37. T. Bynum, 'Computer and Information Ethics', in Zalta (ed.), *Stanford Encyclopedia of Philosophy*, 2008, at http://plato.stanford.edu/entries/ethics-computer [accessed 20 May 2013].

38. K. Mainzer, *Thinking in Complexity: The Computational Dynamics of Matter, Mind, and Mankind* (Berlin and New York: Springer, 2007).

39. H. Haken, *Synergetics: An Introduction: Nonequilibrium Phase Transitions and Self-Organization in Physics, Chemistry and Biology* (Berlin and New York: Springer, 1983); H. Haken, *Synergetic Computers and Cognition* (Berlin and New York: Springer, 2004); L. Steels, 'Semiotic Dynamics for Embodied Agents', *IEEE Intelligent Systems*, 21:3 (2006), pp. 32–8 and Mainzer, *Thinking in Complexity*.

40. G. Beni and J. Wang, 'Swarm Intelligence in Cellular Robotic Systems', *Proceedings of NATO Advanced Workshop on Robots and Biological Systems*, 102 (1989); M. Dorigo et al., 'Evolving Self-Organizing Behaviors for a Swarm-Bot', *Autonomous Robots*, 17:2 (2004), pp. 223–45.

41. G. Dodig-Crnkovic, 'Epistemology Naturalized: The Info-Computationalist Approach', *APA Newsletter on Philosophy and Computers*, 6:2 (2007), pp. 9–14; G. Dodig-Crnkovic, 'Knowledge Generation as Natural Computation', International Conference on Knowledge Generation, Communication and Management – KGCM 2007, at http://www.idt.mdh.se/~gdc/work/PRIS_InfoComp_GDC.pdf [accessed 29 May 2013].

42. Ryle, *The Concept of Mind*; Polanyi, *Personal Knowledge*; Polanyi, *The Tacit Dimension*.

43. J. Stanley and T. Williamson, 'Knowing How', *Journal of Philosophy*, 98:8 (2001), pp. 411–44.

44. A. Noë, 'Against Intellectualism', *Analysis*, 65:4 (2005), pp. 278–90.

45. Shannon and Weaver, *The Mathematical Theory of Communication*.

46. L. Qvortrup, 'The Controversy over the Concept of Information: An Overview and a Selected and Annotated Bibliography', *Cybernetics & Human Knowing*, 14 (1993), pp. 3–24.

47. S. Ott, *Information. Zur Genese und Anwendung eines Begriffs* (Konstanz: UVK Verlagsgesellschaft, 2004).

48. Gruber, 'A Translation Approach to Portable Ontology Specifications'.

49. Smith, 'Beyond Concepts'.

50. J. Seibt, *Process Theories: Cross-Disciplinary Studies in Dynamic Categories* (Dordrecht and London: Kluwer, 2003).

51. I. Nonaka and V. Peltokorpi, 'Objectivity and Subjectivity in Knowledge Management: A Review of 20 Top Articles', *Knowledge and Process Management*, 13:2 (2006), pp. 73–82.

52. E. L. Gettier, 'Is Justified True Belief Knowledge?' *Analysis*, 23:6 (1963), pp. 121–3.

53. Ryle, *The Concept of Mind*; Polanyi, *Personal Knowledge*; Polanyi, *The Tacit Dimension*.

54. M. Minsky, 'A Framework for Representing Knowledge', AI Laboratory, MIT, June 1974, at https://dspace.mit.edu/bitstream/handle/1721.1/6089/AIM-306.pdf [accessed 18 November 2013].

55. N. Chomsky, *Syntactic Structures* (1957; Berlin and New York: Walter de Gruyter, 2002).

56. J. Sneed, *The Logical Structure of Mathematical Physics*, 2nd edn (Dordrecht: Reidel Publishing Company, 1979); C. U. Moulines, 'Introduction: Structuralism as a Program for Modelling Theoretical Science', *Synthese*, 130:1 (2002), pp. 1–11.

57. Heidegger, *Being and Time*.

58. C. Strauss and N. Quinn, *A Cognitive Theory of Cultural Meaning* (Cambridge: Cambridge University Press, 1997).

59. D. J. Saab and U. V. Riss, 'Logic and Abstraction as Capabilities of the Mind', in J. Vallverdù (ed.), *Thinking Machines and the Philosophy of Computer Science: Concepts and Principles* (Hershey, PA: Information Science Publishing, 2010); D. J. Saab and U. V. Riss, 'Information as Ontologization', *Journal of the American Society for Information Science and Technology*, 2:11 (2011), pp. 2236–46.

60. R. Hagengruber and U. V. Riss, 'Knowledge in Action', in G. Dodig Crnkovic and S. Stuart (eds), *Computation, Information, Cognition: The Nexus and the Liminal* (Newcastle upon Tyne: Cambridge Scholars Publishing, 2007), pp. 134–47.

61. P. Janich, *Logish-pragmatische Propädeutik: en Grundkurs im philosophischen Reflektieren* (Weilerwist: Velbrück Wissenschaft, 2001).

62. P. Janich, *Erkennen als Handeln: von der konstruktiven Wissenschaftstheorie zur Erkenntnistheorie. Jenaer philosophische Vorträge und Studien* (Erlangen, Germany: Palm & Enke, 1993); T. Metzinger and V. Gallese, 'The Emergence of a Shared Action Ontology: Building Blocks for a Theory', *Consciousness and Cognition*, 12:4 (2003), pp. 549–71.

63. G. H. von Wright and G. Meggle, *Normen, Werte und Handlungen* (Frankfurt: Suhrkamp, 1994); G. Meggle, *Analytische Handlungstheorie* (Frankfurt: Suhrkamp, 1977); D. M. Wolpert, K. Doya and M. Kawato, 'A Unifying Computational Framework for Motor Control and Social Interaction', *Philosophical Transactions of the Royal Society London B*, 358 (2003), pp. 593–602.

64. P. Janich, 'Informationsbegriff und methodisch-kulturalistische Philosophie', *Ethik und Sozialwissenschaften*, 9:2 (1998), pp. 169–268; G. Ropohl, 'Der Informationsbegriff im Kulturstreit', *Ethik und Sozialwissenschaften. Streitforum für Erwägungskultur*, 12:1 (2001), pp. 3–14.

65. U. V. Riss, 'Knowledge, Action, and Context: Impact on Knowledge Management', in K.-D. Althoff et al. (eds), *Professional Knowledge Management, Lecture Notes in Artificial Intelligence* (Berlin: Springer, 2005), pp. 598–608.

66. E. Craig, *Knowledge and the State of Nature: An Essay in Conceptual Synthesis* (Oxford: Oxford University Press, 1999); A. Kern, *Quellen des Wissens* (Frankfurt: Suhrkamp, 2006).

67. Ibid.

68. Hagengruber and Riss, 'Knowledge in Action'.

69. H. Collins and M. Kusch, *The Shape of Actions: What Humans and Machines Can Do* (Cambridge, MA: MIT Press, 1999).

70. B. Wyssusek, M. Schwartz and B. Kremberg, 'The Philosophical Foundation of Conceptual Knowledge – A Sociopragmatic Approach', in G. W. Mineau and G. Stumme (eds), *Supplementary Proceedings of the 9th International Conference on Conceptual Structures*

(ICCS 2001), Stanford University, 2001, at http://ftp.informatik.rwth-aachen.de/Publications/CEUR-WS/Vol-41/Wyssusek.pdf [accessed 19 November 2013].

71. A. Komus and F. Wauch, *Wikimanagement: Was Unternehmen von Social Software und Web 2.0 lernen können* (Munich: Oldenbourg, 2008).

72. T. Pross, 'Grounded Discourse Representation Theory: Toward a Semantics-Pragmatics Interface for Human-machine Collaboration' (PhD dissertation, IMS University of Stuttgart, 2008).

73. D. Davidson, 'Actions, Reasons and Causes', *Journal of Philosophy*, 60 (1963), pp. 695–700.

74. M. Gilbert, *On Social Facts* (London: Routledge, 1989).

75. R. Tuomela, *The Importance of Us: A Philosophical Study of Basic Social Notions* (Stanford, CA: Stanford University Press, 1995).

76. J. R. Searle, *The Construction of Social Reality* (New York: The Free Press, 1995).

77. Dodig-Crnkovic, 'Epistemology Naturalized'; Dodig-Crnkovic, 'Knowledge Generation as Natural Computation'.

78. G. J. Chaitin, 'Epistemology as Information Theory: Alan Turing Lecture Given at E-CAP 2005', in Dodig-Crnkovic and Stuart (eds), *Computation, Information, Cognition*, pp. 2–17.

79. See H. Putnam, 'Models and Reality', in H. Putnam, *Realism and Reason: Philosophical Papers Volume 3* (Cambridge: Cambridge University Press, 1980), pp. 1–25.

1 Floridi, 'The Fourth Revolution in Our Self-Understanding'

1. US Civil Service, 'The Executive Documents, House of Representatives, First Session of the Fiftyfirst Congress 1889–90', *Serial Set*, 2677 (1889).

2. The phenomenon of telepresence or presence at distance is not crucial in many contexts, such as surgery, where remote control and interactions are becoming widespread.

3. *Dead Souls* is a classic novel by the Russian writer Nikolai Gogol (1809–52). Published in 1842, it centres on Chichikov (the main character) and the people he encounters. The expression 'dead souls' has a twofold meaning. On the one hand, it refers to the fact that, until 1861, landowners in the Russian Empire were entitled to own serfs to farm their land. Serfs were like slaves: they could be bought, sold or mortgaged, and were counted in terms of 'souls'. 'Dead souls' are serfs still accounted for in property registers. On the other hand, 'dead souls' also refers to the characters in the novel, insofar as they have become fake individuals.

4. A. Van Duyn and R. Waters, 'Google in $900m Ad Deal with Myspace', *Financial Times* (8 August 2006), at http://www.ft.com/intl/cms/s/2/17e8e67e-2660-11db-afa1-0000779e2340.html#axzz2k2Sw4OJV [accessed 7 March 2014].

5. N. V. Gogol, *Dead Souls* (New York: New American Library, 1961).

6. S. Raice, 'Facebook Sets Historic IPO', *Wall Street Journal* (2 February 2012), at http://online.wsj.com/news/articles/SB10001424052970204879004577110780078310366 [accessed 7 March 2014].

7. I was not the only one to be astonished; see P. Cohan, 'Yahoo's Tumblr Buy Fails Four Tests of a Successful Acquisition', *Forbes* (20 May 2013), at http://www.forbes.com/sites/petercohan/2013/05/20/yahoos-tumblr-buy-fails-4-tests-of-a-successful-acquisition/ [accessed 7 March 2014].

8. E. Steel, 'Companies Scramble for Consumer Data', *Financial Times* (12 June 2013), at http://www.ft.com/intl/cms/s/0/f0b6edc0-d342-11e2-b3ff-00144feab7de. html#axzz2k2Sw4OJV [accessed 7 March 2014].

2 Krebs, 'Information Transfer as a Metaphor'

1. L. Floridi, 'What is the Philosophy of Information?', in T. W. Bynum and J. H. Moor (eds), *CyberPhilosophy: The Intersection of Philosophy and Computing*, Special Issue of *Metaphilosophy*, 33:1–2 (2002), pp. 117–38.
2. R. Capurro, 'On the Genealogy of Information', in K. Kornwachs and K. Jacoby (eds), *Information: New Questions to a Multidisciplinary Concept* (Berlin: Akademie Verlag, 1996), pp. 259–70.
3. C. Zins, 'Conceptual Approaches for Defining Data, Information, and Knowledge', *Journal of the American Society for Information Science and Technology*, 58:4 (2007), pp. 479–93.
4. C. E. Shannon, *The Lattice Theory of Information* (1950). Reprinted in N. J. A. Sloane and A. D. Wyner (eds), *Claude E. Shannon: Collected Papers* (New York: IEEE Press, 1993).
5. R. Capurro, 'Towards an Ontological Foundation of Information Ethics', *Ethics and Information Technology*, 8:4 (2006), pp. 175–86.
6. R. Millikan, *On Clear and Confused Ideas: An Essay about Substance Concepts* (Cambridge: Cambridge University Press, 2000).
7. C. E. Shannon and W. Weaver, *The Mathematical Theory of Communication* (1948; Urbana, IL: University of Illinois Press, 1999).
8. F. Dretske, *Knowledge and the Flow of Information* (Stanford, CA: CSLI Publication, 1999).
9. D. Davidson, *Truth, Language, and History* (Oxford: Oxford University Press, 2005).
10. Ibid., p. 128.
11. Y. Gunther (ed.), *Essays on Nonconceptual Content* (Cambridge, MA: MIT Press, 2003).
12. G. Peter and G. Preyer (eds), *Contextualism in Philosophy: Knowledge, Meaning, and Truth* (Oxford: Oxford University Press, 2005).
13. S. Soames, *Beyond Rigidity: The Unfinished Semantic Agenda of Naming and Necessity* (Oxford: Oxford University Press, 2002).
14. J. Fodor, *The Mind Doesn't Work that Way: The Scope and Limits of Computational Psychology* (Cambridge, MA: MIT Press, 2001), p. 28.
15. W. Hofkirchner (ed.), *The Quest for a Unified Theory of Information: Proceedings of the Second International Conference on the Foundations of Information Science* (Amsterdam: Gordon & Breach, 1998).
16. M. Esfeld, 'Holism and Analytic Philosophy', *Mind*, new series, 107 (1998), pp. 365–80.
17. D. Davidson, *Inquiries into Truth and Interpretation* (Oxford: Oxford University Press, 1984).
18. J. Fodor and E. Lepore (eds), *Holism: A Consumer Update* (Amsterdam: Rodopi, 1993).
19. L. Wittgenstein, *Philosophische Untersuchungen*, ed. J. Schulte (1953; Frankfurt: Suhrkamp Verlag, 1982).
20. G. W. Bertram, D. Lauer, J. Liptow and M. Seel, *In der Welt der Sprache. Konsequenzen des semantischen Holismus* (Frankfurt: Suhrkamp, 2008).
21. H. Capellen and E. Lepore, *Insensitive Semantics* (Oxford: Basil Blackwell, 2005).

22. D. Kaplan, 'On the Logic of Demonstratives', *Journal of Philosophical Logic*, 8:1 (1979), pp. 81–98.

23. D. Braddon-Mitchell and F. Jackson, *Philosophy of Mind and Cognition: An Introduction* (Oxford: Blackwell, 2008).

24. H. Capellen and E. Lepore, 'A Tall Tale: In Defense of Semantic Minimalism and Speech Act Pluralism', in Peter and Preyer (eds), *Contextualism in Philosophy*, pp. 197–220.

25. P. Pagin, 'Compositionality and Context', in Peter and Preyer (eds), *Contextualism in Philosophy*, pp. 303–48.

26. M. Glanzberg, 'Presuppositions, Truth Values, and Expressing Propositions', in Peter and Preyer (eds), *Contextualism in Philosophy*, pp. 349–98.

27. Ibid.

28. Ibid.

29. F. Recanati, *Literal Meaning* (Cambridge: Cambridge University Press, 2004).

30. P. E. Griffiths, 'Genetic Information: A Metaphor in Search of a Theory', *Philosophy of Science*, 68:3 (2001), pp. 394–412.

31. A. Noë, *Out of Our Heads: Why You Are Not Your Brain, and Other Lessons from the Biology of Consciousness* (New York: Hill and Wang, 2009).

32. J. Fodor, 'Information and Representation', in P. P. Hanson (ed.), *Information, Language, and Cognition* (Vancouver: UBC Press, 1990), pp. 175–190.

33. L. Floridi, 'Outline of a Theory of Strongly Semantic Information', *Minds and Machines*, 14:2 (2004), pp. 197–222.

34. Pagin, 'Compositionality and Context'.

35. M. Black, *Models and Metaphors* (Ithaca, NY: Cornell University Press, 1962).

36. M. Hesse, *Revolutions and Reconstructions in the Philosophy of Science* (Brighton: Harvester Press, 1980).

37. D. Dennett, 'Intuition Pumps', in J. Brockman (ed.), *The Third Culture* (New York: Simon & Schuster, 1995), pp. 180–97.

38. H. J. Schneider, 'Metaphorically Created Objects: "Real" or "Only Linguistic"?', in B. Debatin, T. R. Jackson and D. Steuer (eds), *Metaphor and Rational Discourse* (Tübingen: Niemeyer, 1997), pp. 91–100.

39. Fodor, *The Mind Doesn't Work that Way*.

40. Floridi, 'What is the Philosophy of Information?'.

41. J. McDowell, *Mind and World* (Cambridge, MA: Harvard University Press, 1996).

42. J. Habermas, *On the Pragmatics of Communication* (Cambridge, MA: MIT Press, 1998); R. Brandom, *Making it Explicit: Reasoning, Representing, and Discursive Commitment* (Cambridge, MA: Harvard University Press, 1998).

43. L. Floridi, *Philosophy and Computing: An Introduction* (London: Routledge, 1999).

44. J. Schaffer, 'Knowing the Answer', *Philosophy and Phenomenological Research*, 75:2 (2007), pp. 383–403.

45. M. Frické, 'The Knowledge Pyramid: A Critique of the DIKW Hierarchy', *Journal of Information Science*, 35:2 (2009), pp. 131–42.

3 Voigt, 'With Aristotle towards a Differentiated Concept of Information?'

1. See U. Voigt, *Aristoteles und die Informationsbegriffe. Eine antike Lösung für ein aktuelles Problem?* (Würzburg: Ergon, 2008).

2.   R. Capurro et al., 'Is a Unified Theory of Information Feasible? A Trialogue', *World Futures General Evolution Studies*, 13 (1999), pp. 9–30; P. Fleissner and W. Hofkirchner, 'Informatio Revisited. Wider den dinglichen Informationsbegriff', *Informatik Forum*, 8 (1995), pp. 126–31.

3.   S. Ott, *Information. Zur Genese und Anwendung eines Begriffs* (Konstanz: UVK Verlagsgesellschaft, 2004).

4.   G. Bateson, *Steps to an Ecology of Mind: Collected Essays in Anthropology, Psychiatry, Evolution, and Epistemology* (Chicago, IL: University of Chicago Press, 1972).

5.   Ott, *Information. Zur Genese und Anwendung eines Begriffs*.

6.   L. Qvortrup, 'The Controversy over the Concept of Information: An Overview and a Selected and Annotated Bibliography', *Cybernetics & Human Knowing*, 1 (1993), pp. 3–24.

7.   L. W. Rosenfield, *Aristotle and Information Theory: A Comparison of the Influence of Causal Assumptions on Two Theories of Communication* (The Hague: Mouton, 1971).

8.   D. B. Claus, *Toward the Soul: An Inquiry into the Meaning of RLPZ before Plato* (New Haven, CT: Yale University Press, 1981).

9.   Aristotle, *De Anima* (New York: Cosimo, 2008), I.1.402a6–7.

10. Ibid., II.413a22–5; II.3.414a29–32. See also G. B. Matthews, 'De Anima 2. 2–4 and the Meaning of Life', in M. Nussbaum and A. O. Rorty (eds), *Essays on Aristotle's De Anima* (Oxford: Oxford University Press, 1992), pp. 185–93.

11. Aristotle, *De Anima*, II.3.414b20–415a13.

12. U. Voigt, 'Wozu brauchte Aristoteles den Dualismus? Oder: Warum sich der aktive Geist nicht naturalisieren lässt', in B. Niederbacher and E. Runggaldier (eds), *Die menschliche Seele: Brauchen wir den Dualismus?* (Paderborn: Ontos, 2006), pp. 117–52.

13. Aristotle, *De Anima*, II.3.415a8–13.

14. Ibid., I.4.408b18–29; II.2.413b24–7.

15. Ibid., III.5.430a10–25. See also U. Voigt, 'Wozu brauchte Aristoteles den Dualismus?'.

16. U. Voigt, 'Von Seelen, Figuren und Seeleuten. Zur Einheit und Vielfalt des Begriffs des Lebens (TZ) bei Aristoteles', in S. Föllinger (ed.), *Was ist 'Leben'? Aristoteles' Anschauungen zur Entstehung und Funktionsweise von 'Leben'* (Stuttgart: Steiner, 2010), pp. 17–33.

4 Fuchs-Kittowski, 'The Influence of Philosophy on the Understanding of Computing and Information'

1.   H. Zemanek, *Was ist Informatik?* (Berlin: Springer, 1972).

2.   K. Fuchs-Kittowski, 'Grundlinien des Einsatzes der modernen Informations- und Kommunikationstechnologien in der DDR. Wechsel der Sichtweisen zu einer am Menschen orientieren Informationssystemgestaltun', in F. Naumann and G. Schade (eds), *Informatik in der DDR – eine Bilanz* (Bonn: Gesellschaft für Informatik, 2006), pp. 55–70; K. Fuchs-Kittowski, 'Orientierungen der Informatik in der DDR – Zur Herausbildung von Sichtweiosen für die Gestaltung automatenunterstützter Informationssysteme und zum Ringen um eine sozial orientierte Informatik', in Naumann and Schade (eds), *Informatik in der DDR*, pp. 392–421.

3.   C. Floyd, 'Outline of a Paradigm Change in Software Engineering', in G. Bjerges, P. Ehn and M. Kyng (eds), *Computer and Democracy: A Scandinavian Challenge* (Alderhot: Avebury, 1989), pp. 192–210 and R. Keil-Slawik, 'Artefacts in Software Design', in

C. Floyd et al. (eds), *Software Development and Reality Construction* (Berlin: Springer, 1992), pp. 168–88.

4. K. Fuchs-Kittowski and B. Wenzlaff, 'Integrative Participation – A Challenge to the Development of Informatics', in P. Docherty et al. (eds), *System Design for Human Development and Productivity: Participation and Beyond* (Amsterdam: North Holland, 1987), pp. 3–17.

5. E. von Glasersfeld, 'Declaration of the American Society for Cybernetics', in E. von Glasersfeld, *Radical Constructivism* (London: Falmer, 1995), pp. 146–60, on p. 146.

6. Zemanek, *Was ist Informatik?*

7. C. Floyd et al. (eds), *Software Development and Reality Construction* (Berlin: Springer 1992).

8. H. L. Dreyfus, *What Computers Can't Do: The Limits of Artificial Intelligence* (New York: Harper & Row, 1979).

9. T. Winograd and F. Flores, *Understanding Computer and Cognition: A New Foundation for Design* (Norwood, NJ: Ablex, 1986).

10. A. Raeithel, 'Activity Theory as a Foundation for Design', in C. Floyd et al. (eds), *Software Development and Reality Construction* (Berlin: Springer, 1992), pp. 391–415, on p. 394.

11. M. Hauben and R. Hauben, *Netizens: On the History and the Impact of Usenet and the Internet* (IEEE Computer Society Press: Los Alamitos, 1997); the quote originates from Licklider's unpublished ARPANET Completion Report Draft, 9 September 1977.

12. K. Fuchs-Kittowski et al., 'Man/Computer Communication: A Problem of Linking Semantic and Syntactic Information Processing', in *Workshop on Data Communications* (Laxenburg: International Institute for Applied Systems Analysis – IASA, 1975), pp. 169–88.

13. J. C. R. Licklider, 'Man–Computer Symbiosis', in R. W. Taylor (ed.), *In Memoriam: J. C. R. Licklider*, 7 August 1990, pp. 1–20, at http://memex.org/licklider.pdf [accessed 7 March 2014].

14. J. C. R. Licklider, 'Man–Computer Symbiosis', *IRE Transactions on Human Factors in Electronics*, 1 (1960), pp. 4–11; Licklider, 'Man–Computer Symbiosis', in Taylor (ed.), *In Memoriam: J. C. R. Licklider*.

15. Fuchs-Kittowski et al., 'Man/Computer Communication'.

16. Dreyfus, *What Computers Can't Do: The Limits of Artificial Intelligence*, p. 379.

17. This has been explaind in C. Floyd and S. Ukena, 'On Designing Ontologies for Knowledge Sharing in Communities of Practice', in J. Castro and E. Teniente (eds), *Proceedings of the CAiSE'05 Workshops* (Porto: Faculdade de Engenharia da Universidade do Porto, 2005), pp. 559–69.

18. J. Weizenbaum, *Computer Power and Human Reason* (New York: Freeman, 1976); T. Winograd and F. Flores, *Understanding Computer and Cognition: A New Foundation for Design* (Norwood, NJ: Ablex, 1986).

19. Dreyfus, *What Computers Can't Do: The Limits of Artificial Intelligence* and J. R. Searle, *Speech Acts: An Essay in the Philosophy of Language* (Cambridge: Cambridge University Press, 1969).

20. H. L. Dreyfus and S. E. Dreyfus, *Mind over Machine: The Power of Human Intuition and Expertise in the Era of Computer* (New York: Free Press, 1986).

21. T. R. Gruber, 'A Translation Approach to Portable Ontology Specifications', *Knowledge Acquisition*, 5 (1993), pp. 199–220.

22. R. Mizoguchi, 'Tutorial on Ontological Engineering', Osaka University, Institute of Scientific and Industrial Research, 2004, at http://www.ei.sanken.osaka-u.ac.jp/pub/miz/Part1-pdf2.pdf [accessed 7 March 2014].

23. Raeithel, 'Activity Theory as a Foundation for Design'. Recently, a debate, initiated by G. Merril, started about the interpretation and role of philosophical realism and, in particular, about the type of realism defended by B. Smith. H. Herre believes that an 'integrative realism' overcomes serious weaknesses of the other type of realism. He emphasizes that his approach, though inspired by Brentano's approach, is his own interpretation. See G. H. Merril, 'Ontological Realism: Methodology or Misdirection?', *Applied Ontology*, 5 (2010), pp. 79–108; H. Herre, 'General Formal Ontology: A Foundational Ontology for Conceptual Modelling', in R. Poli and L. Orbst (eds), *Theory and Applications of Ontology. Vol. 2* (Dordrecht: Springer, 2010), pp. 297–345.

24. B. Smith, 'The Basic Tools of Formal Ontology', in N. Guarino (ed.), *Formal Ontology in Information Systems* (Amsterdam: IOS Press, 1998), pp. 19–28; B. Smith, 'Ontology', in L. Floridi (ed.), *The Blackwell Guide to the Philosophy of Computing and Information* (Blackwell: Oxford, 2004), pp. 155–66.

25. Smith, 'The Basic Tools of Formal Ontology'.

26. K. Fuchs-Kittowski and W. Bodrow, 'Wissensmanagement für Wertschöpfung und Wissensschaffung – Allgemeine Prozessontologien als theoretisch-methodologische Grundlage', in G. Banse and E.-O. Reher (eds), *Fortschritte bei der Herausbildung der Allgemeinen Technologie*, (Berlin: Leibniz-Sozietät, 2004), vol. 75, pp. 81–104; K. Fuchs-Kittowski and W. Bodrow, 'Aktivitäten als Basis für Meta-Ontologien in Unternehmen – Werkzeuge zum Wissensmanagement und Bausteine einer Allgemeinen Technologie', in G. Banse and E.-O. Reher (eds), *Fortschritte bei der Herausbildung der Allgemeinen Technologie*, (Berlin: Leibniz-Sozietät, 2008), vol. 99, pp. 221–47.

27. H. Sluga, *Gottlob Frege* (London: Routlege & Kegan Paul, 1980).

28. N. Wiener, *Kybernetik – Regelung und Nachrichtenübertragung im Lebewesen und in der Maschine* (Düsseldorf: Econ, 1963).

29. C. E. Shannon and W. Weaver, *The Mathematical Theory of Communication* (Urbana, IL: University of Illinois Press, 1949).

30. L. Szilard, 'On the Reduction of Entropy in a Thermodynamic System by the Interference of Intelligent Beings', *Z. Physik*, 53 (1929), pp. 840–56. L. Brillouin, 'Life, Thermodynamics, and Cybernetics', *American Scientist*, 37:4 (1949), pp. 554–68. N. Wiener, *Cybernetics or Control and Communication in the Animal and the Machine* (New York: Wiley, 1948).

31. K. Fuchs-Kittowski, *Probleme des Determinismus und der Kybernetik in der molekularen Biologie – Tatsachen und Hypothesen zum Verhältnis von technischem Automaten zum lebenden Organismus* (Jena: VEB G. Fischer, 1976).

32. W. M. Elsasser, *The Physical Foundation of Biology: An Analytical Study* (London: Pergamon Press, 1958).

33. F. J. Varela, *Kognitionswoissemnschaft – Kognitionstechnik – Eine Skizze aktueller Perspektiven*; (Frankfurt: Suhrkamp Verlag, 1990), dust-jacket text.

34. K. Fuchs-Kittowski, 'Information und Biologie: Informationsentstehung – eine neue Kategorie für eine Theorie der Biologie, Biochemie – ein Katalysator der Biowissenschaften', *Sitzungsberichte der Leibniz-Sozietät*, 22:3 (1998), pp. 5–17; K. Fuchs-Kittowski and H. A. Rosenthal, 'Selbstorganisation, Information und Evolution – Zur Kreativität der lebenden Natur', in N. Fenzl, W. Hofkirchner and G. Stockinger (eds), *Information und Selbstorganisation* (Innsbruck and Vienna: Studien Verlag, 1998), pp. 141–88.

35. P. Fleissner and W. Hofkirchner, 'Emergent Information: Towards a Unified Information Theory', *BioSystems, Journal of Biological and Information Processing Sciences*, 38 (1996), pp. 243–48. See also W. Ebeling, J. Freund and F. Schweizer, *Entropie – Information –*

Komplexität (Stuttgart: SFB 230, 1996); K. Fuchs-Kittowski, L. J. Heinrich and A. Rolf, 'Information entsteht in Organisationen – in kreativ lernenden Unternehmen – wissenschaftstheoretische und methodologische Konsequenzen für die Wirtschaftsinformatik', in J. Becker (ed.), *Wirtschaftsinformatik und Wissenschaftstheorie* (Wiesbaden: Gabler, 1999), pp. 329–61; W. Hofkirchner (ed.), *The Quest for a Unified Theory of Information* (Amsterdam: Overseas Publishers Association, 1997).

36. K. Haefner, 'Evolution of Information Processing: Basic Concepts', in K. Haefner (ed.), *Evolution of Information Processing Systems: An Interdisciplinary Approach for a New Understanding of Nature and Society* (Heidelberg and New York: Springer Verlag, 1992), pp. 1–46.

37. C. W. Morris, *Grundlagen der Zeichentheorie* (Munich: Hanser, 1972).

38. *Collected Papers of Charles Saunders Peirce, Vols 1–6*, ed. C. Hartsthorne and P. Weiss, 6 vols (Cambridge, MA: Harvard University Press, 1931–58); *Collected Papers of Charles Saunders Peirce, Vols 7–8*, ed. A. W. Burks, 2 vols (Cambridge, MA: Harvard University Press, 1960).

39. K. Fuchs-Kittowski, 'Information – Neither Matter nor Mind – on the Essence and the Evolutionary Stage Concept of Information', in W. Hofkirchner (ed.), *The Quest for a Unified Theory of Information* (Amsterdam: Gordon & Breach, 1998), pp. 551–70.

40. M. Eigen, 'Selforganization of Matter and the Evolution of Biological Macromolecules', *Naturwissenschaften*, 58:10 (1971), pp. 465–523.

41. Fuchs-Kittowski, *Probleme des Determinismus und der Kybernetik in der molekularen Biologie*; K. Fuchs-Kittowski, 'Reflection on the Essence of Information', in C. Floyd et al. (eds), *Software Development and Reality Construction* (Berlin: Springer, 1992), pp. 416–32; P.-O. Küppers, *Der Ursprung biologischer Information* (Munich: Piper, 1986); and Ebeling, Freund and Schweizer, *Entropie – Information – Komplexität*.

42. Fuchs-Kittowski, Heinrich and Rolf, 'Information entsteht in Organisationen'.

43. H. A. Simon, *The Sciences of the Artificial* (Cambridge, MA: MIT Press, 1969).

44. H.-J. Warnecke, *Fraktale Fabrik – Revolution der Unternehmenskultur* (Berlin: Springer, 1993).

45. M. I. Nurminen, J. Berleur and J. Impagliazzo, 'Social Informatics: An Information Society for All?', in J. Berleur, M. I. Nurminen and J. Impagliazzo (eds), in *Social Informatics: An Information Society for All? In Remembrance of Rob Kling* (New York: Springer, 2006), pp. 1–16; K. Fuchs-Kittowski, 'Quality of Working Life – Knowledge-Intensive Work Processes and Creative Learning Organizations – Information Processing Paradigm Versus Self-Organization Theory', in K. Brunnstein and J. Berleur (eds), *Human Choice and Computers: Issues of Choice and Quality of Life in the Information Society* (Boston, MA and London: Kluwer, 2002), pp. 265–74.

46. J. Berleur, 'Risks and Vulnerability in an Information and Artificial Society', in R. M. Aiken (ed.), *Education and Society: Information Processing 92: Proceedings of the Ifip 12th World Computer Congress Madrid, Spain, 7–11 September 1992*, 3 vols (Amsterdam: North-Holland, 1992), vol. 2, pp. 309–10; K. Brunnstein, 'Perspectives of the Vulnerability of IT-based Society', in Aiken (ed.), *Education and Society: Information Processing 92*, vol. 2, pp. 309–10.

47. K. Fuchs-Kittowski, H.-A. Rosenthal and A. Rosenthal, 'Die Entschlüsselung des Humangenoms – ambivalente Auswirkungen auf Gesellschaft und Wissenschaft', *EWE – Erwägen, Wissen, Ethik*, 16:2 (2005), pp. 149–62, 219–34.

48. B. Smith and R. Hagengruber, 'Zur Zukunft philosophischer Forschung: Die Bedeutung der philosophischen Ontologie in der Informatik. Interview von Ruth Hagengruber mit Barry Smith', *Information Philosophie*, 8 (2003), pp. 132–7.
49. P. T. de Chardin, *Der Mensch im Kosmos* (Munich: Beck, 1969).
50. Zemanek, *Was ist Informatik?*
51. See R. Löther's chapter, 'Die Zukunft der Biosphäre', in J. Bretschneider and H.-G. Eschke (eds), *Humanismus, Menschenwürde und Verantwortung in unserer Zeit* (Neustadt am Rübenberge: Angelika Lenz, 2004), pp. 151–69.
52. K. Fuchs-Kittowski and P. Krüger, 'The Noosphäre Vision of Perre Teilhard de Chardin and Vladimir I. Vernadski in the Perspective of Information and World-Wide Communication 1', in Hofkirchner (ed.), *The Quest for a Unified Theory of Information*, pp. 757–84 and Fuchs-Kittowski, 'Information – Neither Matter nor Mind'.
53. E. Bloch, *Zur Ontologie des Noch-Nicht-Seins: ein Vortrag und zwei Abhandlungen* (Frankfurt: Suhrkamp, 1961.

5 Mainzer, 'The Emergence of Self-Conscious Systems: From Symbolic AI to Embodied Robotics'

1. K. Mainzer, *KI – Künstliche Intelligenz. Grundlagen intelligenter Systeme* (Darmstadt: Wissenschaftliche Buchgesellschaft, 2003).
2. H. L. Dreyfus, *What Computers Can't Do: The Limits of Artificial Intelligence* (New York: Harper & Row, 1979).
3. M. Merleau-Ponty, *Phenomenology of Perception* (London: Routledge & Kegan Paul, 1962).
4. K. Mainzer, *Symmetry and Complexity: The Spirit and Beauty of Nonlinear Science* (Singapore: World Scientific, 2005); K. Mainzer, *Thinking in Complexity: The Computational Dynamics of Matter, Mind, and Mankind* (New York: Springer, 2007).
5. Mainzer, *Thinking in Complexity*.
6. R. Pfeifer and C. Scheier, *Understanding Intelligence*, 5th edn (Cambridge, MA: MIT Press, 2001).
7. W. J. Freeman, 'How and Why Brains Create Sensory Information', *International Journal of Bifurcation and Chaos*, 14 (2005), pp. 515–30.
8. W.-T. Balke and K. Mainzer, 'Knowledge Representation and the Embodied Mind: Towards a Philosophy and Technology of Personalized Informatics', in K.-D. Althoff et al. (eds), *Professional Knowledge Management* (Berlin: Springer, 2005), pp. 586–97.
9. Mainzer, *KI – Künstliche Intelligenz. Grundlagen intelligenter Systeme*; Mainzer, *Thinking in Complexity*.
10. Mainzer, *Symmetry and Complexity*; Mainzer, *Thinking in Complexity*.
11. K. L. Bellman, 'Self-Conscious Modeling', *it – Information Technology*, 4 (2005), pp. 188–94.

6 Zambak, 'Artificial Intelligence as a New Metaphysical Project'

1. Here, by combinatorial explosion, we refer to the problems in which, since the number of possibilities is huge, an optimum result cannot be achieved by a straightforward search. Computer systems provide solutions in a short period of time along with a sizable memory capacity.

2. In the 1950s, Herbert Simon and his colleagues were sure about the coming successes in AI. For instance, Newell and Simon stated that 'It is not [our] aim to surprise or shock you – but the simplest way [we] can summarize is to say that there are now in the world machines that think, that learn and that create. Moreover, their ability to do these things is going to increase rapidly until – in a visible future – the range of problems they can handle will be coextensive with the range to which human mind has been applied'. A. Newell and H. A. Simon, 'Heuristic Problem Solving: The Next Advance in operation Search', *Operation Research*, 6 (1958), pp. 1–10, on p. 8.

3. For a detailed analysis of the *conditions* of the human mind, see A. F. Zambak and R. Vergauwen, 'M-S Models: A New Approach to Models and Simulations in Artificial Intelligence', *Logique et Analyse*, 50 (2007), pp. 179–206.

4. In this essay, the term 'machine intelligence' will be used in a robotic sense that is situated in an agentive position.

5. M. Ringle, 'Philosophy and Artificial Intelligence', in M. Ringle (ed.), *Philosophical Perspectives in Artificial Intelligence* (Sussex: The Harvester Press, 1979), pp. 1–20, on p. 3.

6. K. M. Sayre, 'The Simulation of Epistemic Acts', in Ringle (ed.), *Philosophical Perspectives in Artificial Intelligence*, pp. 139–60.

7. H. Kyburg, 'Epistemology and Computing', in T. W. Bynum and J. H. Moor (eds), *The Digital Phoenix: How Computers Are Changing Philosophy* (Oxford: Blackwell Publishers, 1998), pp. 37–47.

8. L. Darden, 'Anomaly-Driven Theory Redesign: Computational Philosophy of Science Experiments', in Bynum and Moor (eds), *The Digital Phoenix*, pp. 62–78.

9. J. McCarthy and P. J. Hayes, 'Some Philosophical Problems from the Standpoint of Artificial Intelligence', in B. Meltzer, D. Michie and M. Swann (eds), *Machine Intelligence, Vol. 4* (Edinburgh: Edinburgh University Press, 1969), pp. 463–502.

10. K. M. Ford, C. Glymour and P. J. Hayes (eds), *Android Epistemology* (Cambridge, MA: The MIT Press, 1995), p. 3.

11. M. Boden, *Artificial Intelligence and Natural Man* (London: The MIT Press, 1987).

12. Ford et al. describe the term 'android epistemology' by saying that 'Android epistemology is the business of exploring the space of possible machines and their capacities for knowledge, belief, attitudes, desires, and action in accordance with their mental states. Parts of the enterprise naturally concern the function of machine components and processes; parts of it concern characterizations of machine behavior; parts of it, the proofs of boundaries on what is possible behavior for machine sharing some feature; and parts of android epistemology inevitably involve trying to decide what to say and what not to say about machine parts, states and processes using the psychological terms we use for one another, and similarly, what engineering vocabulary is useful in psychological descriptions of biological machines. Wherever psychologists propose that human thought or belief or desire are generated by some machine process, psychology is android epistemology. Humans are just a special case; only gods are left out of android epistemology.' Ford, Glymour and Hayes (eds), *Android Epistemology*, p. xi.

13. Steinhart considers *digital metaphysics* to be an explanatory methodology for the mental and natural phenomena: 'According to digital metaphysics, physical phenomena emerge from the interactions of monads running programs. One of the major virtues of such computational explications of physical phenomena is that they offer *procedurally effective explanations*, rather than mere descriptions. These explanations state what nature is *doing*.' E. Steinhart, 'Digital Metaphysics', in Bynum and Moor (eds), *The Digital Phoenix*, pp. 117–34.

14. Ibid.
15. M. Boden, 'Methodological Links between AI and Other Disciplines', in S. B. Torrance (ed.), *The Mind and the Machine: Philosophical Aspects of Artificial Inteligence* (Sussex: Ellis Horwood Limited, 1984), pp. 125–32.
16. P. McCorduck, *Machines Who Think: A Personal Inquiry into the History and Prospects of Artificial Intelligence* (San Francisco, CA: W. H. Freeman and Company, 1979).
17. Ringle, 'Philosophy and Artificial Intelligence'.
18. J. L. Pollock, 'Procedural Epistemology', in Bynum and Moor (eds), *The Digital Phoenix*, pp. 17–36.
19. A. Sloman, *The Computer Revolution in Philosophy: Philosophy, Science and Models of Mind* (Sussex: The Harvester Press, 1978).
20. P. Thagard, 'Computation and the Philosophy of Science', in Bynum and Moor (eds), *The Digital Phoenix*, pp. 48–61.
21. J. Doyle, 'The Foundations of Psychology: A Logico-Computational Inquiry into the Concept of Mind', in R. Cummins and J. Pollock (eds), *Philosophy and AI: Essays at the Interface* (Cambridge, MA: MIT Press, 1991), pp. 39–77.
22. Sloman, *The Computer Revolution in Philosophy*.
23. It is a scholastic adage that *actio est perfectio agentis*. However, every action cannot be considered in an agentive manner. Agency is not only a transitive, temporary, momentary and interim action, but has all the characteristics of an embodied action. Agentive action is the source of human embodiment which determines the way we think and behave. Arbib considers human embodiment in an agentive manner in order to overcome certain difficulties in AI: 'I will accept that AI is limited now, but will counter certain philosophical arguments that AI is limited in principle. Nonetheless, when we counterpose intelligence in abstracto with what is specifically human about the way in which we are intelligent, I shall argue that we must take into account the way in which we are embodied, both in having human bodies and in being participants in human society.' M. A. Arbib, *In the Search of the Person: Philosophical Explorations in Cognitive Science* (Amherst, MA: The University of Massachusetts Press, 1985), p. 29.

7 Smith, 'The Relevance of Philosophical Ontology to Information and Computer Science'

1. R. Ingarden, *Time and Modes of Being*, trans. H. Michejda (Springfield, IL: Charles Thomas, 1964).
2. See P. Ørstrøm, J. Andersen and H. Schärfe, 'What Has Happened to Ontology?', in F. Dau, M.-L. Mugnier and G. Stumme (eds), *Conceptual Structures: Common Semantics for Sharing Knowledge, Lecture Notes in Computer Science 3596* (Berlin, Heidelberg: Springer, 2005), pp. 425–38; P. Øhrstrøm, S. Uckelman and H. Schärfe, 'Historical and Conceptual Foundations of Diagrammatical Ontology', in U. Priss, S. Polovina and R. Hill (eds), *Conceptual Structures: Knowledge Architectures for Smart Applications, Proceedings of the 15th International Conference on Conceptual Structures* (Berlin: Springer, 2007), pp. 374–86.
3. Ingarden, *Time and Modes of Being*.
4. L. Johansson, *Ontological Investigations: An Inquiry into the Categories of Nature, Man and Society* (New York and London: Routledge, 1989).

5. R. Chisholm, *A Realistic Theory of Categories: An Essay on Ontology* (Cambridge: Cambridge University Press, 1996).

6. E. Lowe, *The Four-Category Ontology: A Metaphysical Foundation for Natural Science* (Oxford: Oxford University Press, 2006).

7. W. v. O. Quine, 'On What There is', reprinted in W. v. O. Quine, *From a Logical Point of View* (New York: Harper & Row, 1953), pp. 1–19.

8. J. McCarthy, 'Circumscription: A Form of Non-Monotonic Reasoning', *Artificial Intelligence*, 13:5 (1980), pp. 27–39.

9. P. Hayes, 'The Second Naive Physics Manifesto', in J. Hobbs and R. Moore (eds), *Formal Theories of the Common-Sense World* (Norwood: Ablex, 1985), pp. 1–36; P. Hayes, 'Naïve Physics I: Ontology for Liquids', in Hobbs and Moore (eds), *Formal Theories of the Common-Sense World*, pp. 71–108.

10. M. Ashburner et al., 'Gene Ontology: Tool for the Unification of Biology', *Nature Genetics*, 25 (2000), pp. 25–9.

11. Hayes, 'The Second Naive Physics Manifesto'.

12. S. Schulz et al., 'Strengths and Limitations of Formal Ontologies in the Biomedical Domain', *Electronic Journal of Communication, Information and Innovation in Health*, Special Issue on Ontologies, Semantic Webs and Health, 3:1 (2009), pp. 31–45.

13. B. Smith, 'Ontology (Science)', in C. Eschenbach and M. Gruninger (eds), *Formal Ontology in Information Systems*: *Proceedings of the Fifth International Conference* (Amsterdam: IOS Press, 2008), pp. 21–35.

14. B. Smith et al., 'The OBO Foundry: Coordinated Evolution of Ontologies to Support Biomedical Data Integration', *Nature Biotechnology*, 25:11 (2007), pp. 1251–5.

15. See B. Smith et al., 'Relations in Biomedical Ontologies', *Genome Biology*, 6:5 (2005), R46.

16. M. Genesereth and N. Nilsson, *Logical Foundation of Artificial Intelligence* (Los Altos, CA: Morgan Kaufmann, 1987).

17. T. R. Gruber, 'A Translation Approach to Portable Ontology Specifications', *Knowledge Acquisition*, 5 (1993), pp. 199–220.

18. T. R. Gruber, 'Toward Principles for the Design of Ontologies Used for Knowledge Sharing', *International Journal of Human and Computer Studies*, 43:5–6 (1995), pp. 907–28, on p. 908.

19. Smith et al., 'The OBO Foundry: Coordinated Evolution of Ontologies to Support Biomedical Data Integration'.

20. Portions of this essay are based on material taken from my chapter 'Ontology', in L. Floridi (ed.), *Blackwell Guide to the Philosophy of Computing and Information* (Oxford: Blackwell, 2004), pp. 155–66, and from H. Stenzhorn et al., 'Adapting Clinical Ontologies in Real-World Environments', *Journal of Universal Computer Science*, 14:22 (2008), pp. 3767–80.

8 Kohne, 'Ontology, its Origins and its Meaning in Information Science'

1. See W. v. O. Quine, 'On What There is', *Review of Metaphysics*, 2 (1948), pp. 21–38.

2. For example T. Gruber, 'Ontology', in L. Liu and M. T. Özsu (eds), *Encyclopedia of Database Systems* (Berlin: Springer, 2009); B. Smith, 'Ontology', in L. Floridi (ed.), *The Blackwell Guide to the Philosophy of Computing and Information* (Oxford: Black-

well, 2004), pp. 155–66; W. Hesse, 'Ontologien', *Informatik Spektrum*, 25 (2002), pp. 477–80.

3. E. Vollrath, 'Die Gliederung der Metaphysik in eine Metaphysica Generalis und eine Metaphysica Specialis', *Zeitschrift für philosophische Forschung*, 16:2 (1962), pp. 258–84; M. J. Loux, *Metaphysics* (London and NY: Routledge, 1998); Smith, 'Ontology'.

4. Quine, 'On What There is'; Loux, *Metaphysics*.

5. M. J. Loux, *Metaphysics*.

6. Ibid.

7. Ibid., p. 9.

8. Ibid., p. 8.

9. For more about realism see J. Kohne, *Drei Variationen über Ähnlichkeit* (Hildesheim, Zurich and NY: Olms, 2005); J. Kohne, 'Ähnlichkeit', *Conjectura*, 10:2 (2005), pp. 7–41.

10. For more about nominalism see ibid.

11. Loux, *Metaphysics*, p. 8.

12. D. M. Armstrong, *Universals: An Opinionated Introduction* (Boulder, CO: Westview, 1989).

13. Loux, *Metaphysics*, p. 9.

14. For example Gruber, 'Ontology'; Smith, 'Ontology'; Hesse, 'Ontologien'.

15. For example ibid.

16. For example Gruber, 'Ontology'.

17. Ibid.

9 Jaskolla and Rugel, 'Smart Questions: Steps towards an Ontology of Questions and Answers'

1. See B. Smith, F. Neuhaus and P. Grenon, 'A Formal Theory of Substances, Qualities, and Universals', 2004, at http://ontology.buffalo.edu/bfo/SQU.pdf [accessed 10 June 2013]; B. Smith, 'Beyond Concepts: Ontology as Reality Representation', in A. C. Varzi and L. Vieu (eds), *Formal Ontology in Information Systems: Proceedings of the Third International Conference (FOIS-2004), Frontiers in Artificial Intelligence and Applications* (Amsterdam: IOS Press, 2004), pp. 73–84; P. Grenon, 'BFO in a Nutshell: A Bi-Categorial Axiomatization of BFO and Comparison with DOLCE', *IFOMIS REPORTS*, 6 (2003), at http://www.ifomis.org/Research/IFOMISReports/IFOMIS%20 Report%2006_2003.pdf [accessed 10 June 2013].

2. See Aristotle, *Categories*, ed. J. Barnes (Princeton, NJ: Princeton University Press, 1998); R. Chisholm, *A Realistic Theory of Categories. An Essay on Ontology* (Cambridge: Cambridge University Press, 1996).

3. See H. Schuman, *Questions and Answers in Attitude Surveys: Experiments on Question Form, Wording, and Context*, 3rd edn (Thousand Oaks, CA: SAGE Publications, 1997).

4. See H. Toutenburg and C. Heumann, *Deskriptive Statistik. Eine Einführung in Methoden und Anwendungen mit SPSS* (Berlin: Springer, 2006); H. Komrey, *Empirische Sozialforschung* (Stuttgart: UTB, 2006).

5. B. Smith and P. Grenon, 'The Cornucopia of Formal-Ontological Relations', *Dialectica*, 58:3 (2004), pp. 279–96.

6. See Toutenburg and Heumann, *Deskriptive Statistik*.

7. T. Knieper, *Statistik. Eine Einführung für Kommunikationsberufe, Vol. 5* (Munich: UVK, Uni-Papers, 2004).

10 Bringsjord, Clark and Taylor, 'Sophisticated Knowledge Representation and Reasoning Requires Philosophy'

1. R. J. Brachman and H. J. Levesque, *Knowledge Representation and Reasoning* (San Francisco, CA: Elsevier, 2004).

2. S. Bringsjord and Y. Yang, 'Representations Using Formal Logics', in L. Nadel (ed.), *Encyclopedia of Cognitive Science*, 4 vols (London: Nature Publishing Group, 2003), vol. 3, pp. 940–50.

3. For example, see R. Sun and S. Bringsjord, 'Cognitive Systems and Cognitive Architectures', in B. W. Wah (ed.), *The Wiley Encyclopedia of Computer Science and Engineering*, 5 vols (New York: Wiley, 2009), vol. 1, pp. 4200–88; or S. Bringsjord, 'The Logicist Manifesto: At Long Last Let Logic-Based AI Become a Field Unto Itself', *Journal of Applied Logic*, 6:4 (2008), pp. 502–52. The first account is brief, and the second more extensive.

4. Nice coverage of the propositional calculus is provided in J. Barwise and J. Etchemendy, *Language, Proof, and Logic* (New York: Seven Bridges, 1999).

5. For a discussion of the relationship between this concept of intuitive formal validity, provability in a deductive calculus, and the corresponding semantic consequence of $\mathcal{Y}$ following deductively from $\{\varphi_1,...,\varphi_n\}$ provided that there is no model in which all the φ_i are true while $\mathcal{Y}$ is false, see G. Kreisel, 'Informal Rigor and Completeness Proofs', in I. Lakatos (ed.), *Problems in the Philosophy of Mathematics* (Amsterdam: North-Holland, 1967), pp. 138–86.

6. F. B. Fitch, *Symbolic Logic: An Introduction* (New York: Ronald Press, 1952).

7. This is set out in Barwise and Etchemendy, *Language, Proof, and Logic*.

8. K. Arkoudas, 'Athena', at http://www.proofcentral.org/athena [accessed 23 September 2013].

9. The rule for *existential introduction* is expressed below, wherein $\varphi(a)$ is a formula in which a appears as a constant, and $\varphi(a/x)$ is that formula changed only by the replacement of a with the variable x.

$$\frac{\varphi(a)}{\exists x\, \varphi(a/x)}$$

10. For example, Otter, given these two lines as input, produces the following proof:

     --------------- PROOF ---------------

     1 [] -Likes(x,y)|Likes(z,x).

     2 [] -Likes($c1,b).

     3 [] Likes(a,b).

     4 [hyper,3,1] Likes(x,a).

     5 [hyper,4,1] Likes(x,y).

11. For an overview of Semantic Web, see T. Berners-Lee, J. Hendler and O. Lassila, 'The Semantic Web: A New Form of Web Content That is Meaningful to Computers Will Unleash a Revolution of New Possibilities', *Scientific American*, 284:5 (2001), pp. 34–43.

12. F. Baader, D. Calvanese and D. McGuinness (eds), *The Description Logic Handbook: Theory, Implementation and Applications* (Cambridge: Cambridge University Press, 2003).

13. The core theorems can be found in G. S. Boolos, J. P. Burgess and R. C. Jeffrey, *Computability and Logic*, 4th edn (Cambridge: Cambridge University Press, 2003).

14. The Church–Turing thesis states that a function f is effectively computable if and only if f is Turing-computable. For a discussion, see S. Bringsjord and K. Arkoudas, 'On the Provability, Veracity, and AI-Relevance of the Church-Turing Thesis', in A. Olszewski, J. Wolenski and R. Janusz (eds), *Church's Thesis After 70 Years* (Frankfurt: Ontos, 2006), pp. 66–18.

15. Such advanced epistemic logic can be found in K. Arkoudas and S. Bringsjord, 'Metareasoning for Multi-Agent Epistemic Logics', in J. Leite and P. Torroni (eds), *Computational Logic in Multi-Agent Systems: 5th International Workshop, CLIMA V, Lisbon, Portugal, September 29–30, 2004, Revised Selected and Invited Papers* (New York: Springer, 2005), pp. 111–25.

16. There are straightforward algorithms for adapting arbitrary formulas so as to produce such 'front-loaded' versions of them. For example, see Boolos, Burgess and Jeffrey, *Computability and Logic*, 4th edn.

17. S. C. Kleene, *Introduction to Metamathematics* (New York: Van Nostrand, 1952).

18. For example, even book-length presentations of mutation and natural selection routinely fail to provide any formal, rigorous statement of the theory; see, for instance, D. C. Dennett, *Darwin's Dangerous Idea: Evolution and the Meanings of Life* (New York: Simon & Shuster, 1995).

19. Nice coverage is provided in P. Smith, *An Introduction to Gödel's Theorems* (Cambridge: Cambridge University Press, 2007).

20. Where $\varphi(x)$ is a first-order formula with x, and possibly others, as free variables, the universal closure of $\varphi(x)$ binds the variables other than x to universal quantifiers. For example, the universal closure of $B(x, y, z)$ would be $\forall y \forall z\, B(x, y, z)$.

21. A succinct proof of this theorem can be found in H.-D. Ebbinghaus, J. Flum and W. Thomas, *Mathematical Logic* (New York: Springer, 1984).

22. One example is Goodstein's theorem. This theorem, and others in the relevant class, are too complex to cover in the present essay. See P. Smith, *An Introduction to Gödel's Theorems* (Cambridge: Cambridge University Press, 2007).

23. This being an essay on philosophy-empowered KR&R, as opposed to deception and counterdeception, an idealized example is appropriate. We do not have the space to exhibit the power of the brand of KR&R we are championing in connection with real-world deception and counterdeception. Motivated readers can confirm that our techniques are applicable in real-world scenarios by consulting M. Bennett and E. Waltz, *Counterdeception Principles and Applications for National Security* (Norwook: Artech House, 2007).

24. Arkoudas, 'Athena'.

25. Whether the disjunction 'L believes that p is not true or L believes that p is false' is redundant depends on how one formally represents beliefs about propositions. In ΣXX, the formal system we use to define lying precisely, there is no representational difference between believing a proposition to be not true and believing the proposition to be false. However, in other formal systems there may be a representational and logical distinction between the two.

26. Linguistic convention dictates that statements are assertions by default, i.e. when cues to the contrary, such as irony and humour, are absent. The conditions mentioned in the definition of *asserting* are meant to exclude situations where the speaker believes that he/she will be understood as making a non-solemn statement – for example, when the

speaker makes a joke, uses a metaphor, or conveys by other indicator (e.g. a wink or a nod) that he/she is not intending to be taken seriously.

27. R. M. Chisholm and T. D. Feehan, 'The Intent to Deceive', *Journal of Philosophy*, 74:3 (1977), pp. 143–59.

28. For example: (i) '*L* believes that *p* is false' is an expression of a higher-order belief – this belief cannot be attained unless *L* has the concept of something *being false*; (ii) *L*'s beliefs, and *L*'s beliefs about *D*'s beliefs, are both occurrent and defeasible – the latter, defeasibility, indicates that *justifications* ought to be treated as first-class entities within a formal system. See M. H. Clark, 'Cognitive Illusions and the Lying Machine: A Blueprint for Sophistic Mendacity' (PhD dissertation, Rensselaer Polytechnic Institute, Department of Cognitive Science, Troy, 2009).

29. K. Arkoudas and S. Bringsjord, 'Toward Formalizing Common-Sense Psychology: An Analysis of the False-Belief Task', in T.-B. Ho and Z.-H. Zhou (eds), *PRICAI 2008: Trends in Artificial Intelligence, 10th Pacific Rim International Conference on Artificial Intelligence*, Hanoi, Vietnam, 15–19 December 2008 (New York: Springer, 2008), pp. 17–29.

30. We elide the axiom stipulating the temporal order of m_1, m_2, and m_3. Informally, m_1 is prior to m_2 and m_2 is prior to m_3.

31. S. Bringsjord and D. Ferrucci, *Artificial Intelligence and Literary Creativity: Inside the Mind of Brutus, a Storytelling Machine* (Mahwah, NJ: Lawrence Erlbaum, 2000).

32. Some preliminary work is described in S. Bringsjord et al., 'Advanced Synthetic Characters, Evil, and E', in M. Al-Akaidi and A. E. Rhalibi (eds), *Game-On 2005, 6th International Conference on Intelligent Games and Simulation* (Ghent and Zwijnaarde: European Simulation Society, 2005), pp. 31–9.

33. M. S. Peck, *People of the Lie* (New York: Simon & Shuster, 1983).

34. A nice introduction for the more industrious of our readers can be obtained in G. Priest, R. Routley and J. Norman (eds), *Paraconsistent Logic: Essays on the Inconsistent* (Munich: Philosophia, 1989).

35. Sometimes the adjective 'symbolic' is used instead of 'linguistic'.

36. *The Collected Papers of Charles Sanders Peirce*, ed. C. Hartshorne and P. Weiss, 8 vols (Bristol: Thoemmes Press, 1998).

37. C. Glymour, *Thinking Things Through: An Introduction to Philosophical Issues and Achievements* (Cambridge, MA: MIT Press, 1992).

38. J. Barwise and J. Etchemendy, 'Information, Infons, and Inference', in *Situation Theory and its Applications* (Menlo Park, CA: CSLI, 2008), pp. 33–78. The Vivid family is presented in K. Arkoudas and S. Bringsjord, 'Vivid: An AI Framework for Heterogeneous Problem Solving', *Artificial Intelligence*, 173:15 (2009), pp. 1367–405.

11 Andreas, 'On Frames and Theory-Elements of Structuralism'

1. M. Minsky, 'A Framework for Representing Knowledge', AI Laboratory, MIT, June 1974.

2. This methodology was developed by J. Sneed, *The Logical Structure of Mathematical Physics*, 2nd edn (Dordrecht: Reidel Publishing Company, 1979) and W. Balzer, C. U. Moulines and J. Sneed, *An Architectonic for Science: The Structuralist Program* (Dordrecht: Reidel Publishing Company, 1987).

3. Minsky, 'A Framework for Representing Knowledge'.

4. Ibid.

5. Sneed, *The Logical Structure of Mathematical Physics*.
6. T. S. Kuhn, *The Structure of Scientific Revolutions*, 3rd edn (Chicago, IL: University of Chicago Press, 1970).
7. Balzer, Moulines and Sneed, *An Architectonic for Science*.
8. Foundational work of how a complete formalization of claims in structuralism can be accomplished has been done in H. Andreas, *Carnaps Wissenschaftslogik* (Paderborn: Mentis, 2007).
9. There is only a semi-decision procedure available for the validity of first-order formulae; the validity problem for first-order formulae is undecidable. There are theorem provers developed for higher-order logic based on, for example, tableaux methods or resolution.
10. See K. Baclawski and T. Niu, *Ontologies for Bioinformatics* (Cambridge, MA: MIT Press, 2006), p. 66 for discussion of this point.
11. See N. Guarino, 'Formal Ontology in Information Systems', in N. Guarino (ed.), *Formal Ontology in Information Systems: Proceedings of the First International Conference (FOIS-98)* (Amsterdam and Oxford: IOS Press, 1998), pp. 3–15; P. Hitzler et al., *Semantic Web* (Berlin: Springer, 2008).
12. See, for example, J. A. Blake and C. J. Bult, 'Beyond the Data Deluge: Data Integration and Bio-Ontologies', *Journal of Biomedical Informatics*, 39 (2006), pp. 314–20; Z. Lacroix (ed.), *Bioinformatics: Managing Scientific Data* (San Francisco, CA: Morgan Kaufmann, 2003); and L. N. Soldatova and R. D. King, 'An Ontology of Scientific Experiments', *Journal of the Royal Society Interface*, 3:11 (2006), pp. 795–803.
13. These efforts are also motivated by there being a large number of studies in which the structuralist framework has been used for the logical reconstruction of theories in the natural and other sciences, including also the field of biology. Structuralist theory of science is way beyond the discussion of toy examples. The wide-ranging success is documented by two bibliographies: W. Diederich, A. Ibarra and T. Morman, 'Bibliography of Structuralism', *Erkenntnis*, 30 (1989), pp. 387–407; and W. Diederich, A. Ibarra and T. Morman, 'Bibliography of Structuralism II', *Erkenntnis*, 41 (1994), pp. 403–18. Papers have already appeared arguing for and partially realizing an application of the structuralist framework in knowledge representation. See W. Balzer and C. U. Moulines, 'Introduction', in W. Balzer, J. Sneed and C. U. Moulines (eds), *Structuralist Knowledge Representation: Paradigmatic Examples* (Rodopi: Amsterdam, 2000), pp. 8–10; C.-D. Wajnberg et al., 'A Structuralist Approach towards Computational Scientific Discovery', in S. Arikawa and E. Suzuki (eds), *Discovery Science* (Berlin: Springer, 2004), pp. 412–19; and J. Sneed, 'Machine Models for the Growth of Knowledge: Theory Nets in PROLOG', in K. Gavroglu et al. (eds), *Theories of Scientific Change* (Dordrecht: Kluwer, 1989), pp. 245–67.

12 Saab and Fonseca, 'Ontological Complexity and Human Culture'

1. F. Fonseca, 'The Double Role of Ontologies in Information Science Research', *Journal of the American Society for Information Science and Technology*, 56:6 (2007), pp. 786–93.
2. B. Chandrasekaran, J. R. Josephson and V. R. Benjamins, 'What Are Ontologies, and Why Do We Need Them?', *Intelligent Systems and their Applications*, IEEE, 14:1 (1999), pp. 20–6; T. R. Gruber, 'Toward Principles for the Design of Ontologies Used for Knowledge Sharing', in N. Guarino and R. Poli (eds), *Formal Ontology in Conceptual Analysis and Knowledge Representation* (Deventer: Kluwer Academic Publishers, 1993), pp. 907–28; T. R. Gruber and G. R. Olsen, 'An Ontology for Engineering Mathematics',

in J. Doyle, P. Torasso and E. Sandewall (eds), *Fourth International Conference on Principles of Knowledge Representation and Reasoning* (Bonn: Gustav Stresemann Institut, Morgan Kaufmann, 1994), pp. 258–69.

3.   B. Smith, 'Ontology', in L. Floridi (ed.), *The Blackwell Guide to Philosophy of Computing and Information* (Oxford: Blackwell, 2004), pp. 155–66, on pp. 157–8.

4.   T. Winograd and F. Flores, *Understanding Computers and Cognition: A New Foundation for Design* (Boston, MA: Addison-Wesley, 1987).

5.   M. Heidegger, *Being and Time* (New York: Harper & Row, 1927).

6.   R. D'Andrade, *The Development of Cognitive Anthropology* (Cambridge: Cambridge University Press, 1995); C. Strauss and N. Quinn, *A Cognitive Theory of Cultural Meaning* (Cambridge, MA: Cambridge University Press, 1997).

7.   Aristotle, *The Basic Works* (New York: Random House, 1941).

8.   B. Smith, 'Beyond Concepts: Ontology as Reality Representation', in A. Varzi and L. Vieu (eds), *Proceedings of FOIS 2004*, at http://ontology.buffalo.edu/bfo/BeyondConcepts.pdf [accessed 25 September 2013].

9.   Winograd and Flores, *Understanding Computers and Cognition*.

10. Smith, 'Ontology'.

11. F. Keil, *Semantic and Conceptual Development: An Ontological Perspective* (Cambridge, MA: Harvard University Press, 1979); Smith, 'Ontology'; E. S. Spelke, 'Principles of Object Perception', *Cognitive Science*, 14 (1990), pp. 29–56.

12. Chandrasekaran, Josephson and Benjamins, 'What Are Ontologies, and Why Do We Need Them?'.

13. W. Kuśnierczyk, 'Nontological Engineering', in B. Bennett and C. Fellbaum (eds), *Formal Ontology in Information Systems: Proceedings of the Fourth International Conference (FOIS-2006)* (Baltimore, MD: IOS Press, 2006), pp. 39–50.

14. M. K. Smith, C. Welty and D. McGuinness, 'Owl Web Ontology Language Guide 2004', at http://www.w3.org/TR/owl-guide/ [accessed 25 September 2013].

15. A. Gómez-Pérez, O. Corcho and M. Fernández-López, *Ontological Engineering: Advanced Information and Knowledge Processing* (London: Springer, 2004).

16. P. Clark and B. Porter, 'KM – the Knowledge Machine 2.0: Users' Manual', at http://www.cs.utexas.edu/users/mfkb/km/userman.pdf [accessed 25 September 2013].

17. Gómez-Pérez, Corcho and Fernández-López, *Ontological Engineering*.

18. V. K. Chaudhri et al., 'Open Knowledge Base Connectivity 2.0.3.', 1998, at ftp://ftp.ksl.stanford.edu/pub/KSL_Reports/KSL-98-06.pdf [accessed 24 September 2013].

19. Gómez-Pérez, Corcho and Fernández-López, *Ontological Engineering*.

20. S. Bechhofer et al., 'Owl Web Ontology Language Reference 2004', at http://www.w3.org/TR/2004/REC-owl-ref-20040210/ [accessed 24 September 2013].

21. Gómez-Pérez, Corcho and Fernández-López, *Ontological Engineering*.

22. Ibid.

23. C. Welty, 'The Ontological Nature of Subject Taxonomies', in N. Guarino (ed.), *Formal Ontology in Information Systems: Proceedings of the First International Conference (FOIS-98)* (Amsterdam: IOS Press, 1998), pp. 317–27.

24. Kuśnierczyk, 'Nontological Engineering', p. 43.

25. Smith, 'Beyond Concepts'.

26. Gruber and Olsen, 'An Ontology for Engineering Mathematics'.

27. Smith, 'Ontology'.

28. Heidegger, *Being and Time*.

29. M. Heidegger, *The Question Concerning Technology and Other Essays* (New York: Harper & Row, 1977).
30. S. Staab et al., 'Emergent Semantics', *IEEE Intelligent Systems*, 17:1 (2002), pp. 78–86.
31. Heidegger, *Being and Time*.
32. D. E. Rumelhart and J. L. McClelland, *Parallel Distributed Processing: Exploration in the Microstructure of Cognition, Vols. 1 & 2, Psychological and Biological Models* (Cambridge, MA: MIT Press, 1986).
33. R. C. Anderson, R. J. Spiro and W. E. Montague (eds), *Schooling and the Acquisition of Knowledge* (Hillsdale: Lawrence Erlbaum, 1984); D'Andrade, *The Development of Cognitive Anthropology*; P. M. Davis, *Cognition and Learning: A Review of the Literature with Reference to Ethnolinguistic Minorities* (Dallas, TX: Summer Institute of Linguistics, 1991); Strauss and Quinn, *A Cognitive Theory of Cultural Meaning*.
34. They have also been referred to variously in the literature as frames, scenes, scenarios, scripts, models and theories.
35. Winograd and Flores, *Understanding Computers and Cognition*.
36. P. DiMaggio, 'Culture and Cognition', *Annual Review of Sociology*, 23 (1997), pp. 263–88.
37. Strauss and Quinn, *A Cognitive Theory of Cultural Meaning*.
38. L. Talmy, 'The Cognitive Culture System', *Monist*, 78:1 (2001), pp. 81–116.
39. D'Andrade, *The Development of Cognitive Anthropology*; Strauss and Quinn, *A Cognitive Theory of Cultural Meaning*.
40. Heidegger, *Being and Time*; J. Recker, 'Developing Ontological Theories for Conceptual Models Using Qualitative Research', *Proceedings of QualIT2005: Challenges for Qualitative Research*, Brisbane, 23–5 November 2005, at http://eprints.qut.edu.au/2880/1/Recker-QualIT2005.pdf [accessed 25 September 2013].
41. F. Fonseca and J. Martin, 'Play as the Way out of the Newspeak-Tower of Babel Dilemma in Data Modeling', *Proceedings of the Twenty-Sixth International Conference on Information Systems: Philosophy and Research Methods in Information Systems (ICIS) 2005* at http://aisel.aisnet.org/icis2005/26/ [accessed 25 September 2013]; Smith, 'Ontology'.
42. D. M. Mark and A. G. Turk, 'Landscape Categories in Yindjibarndi: Ontology, Environment, and Language', in W. Kuhn, M. Worboys and S. Timpf (eds), *Spatial Information Theory: Foundations of Geographic Information Science* (Berlin: Springer, 2003), pp. 28–45; D. J. Saab, Conceptualizing Space: Mapping Schemas as Meaningful Representations' (MA dissertation, Lesley University, 2003).
43. G. Fauconnier and M. Turner, 'Conceptual Integration Networks', *Cognitive Science*, 22:2 (1998), pp. 133–87.
44. Talmy, 'The Cognitive Culture System'.

13 Riss, 'Knowledge and Action between Abstraction and Concretion'

1. P. Drucker, *Post-Capitalist Society* (London: Butterworth Heinemann, 1993); I. Nonaka and H. Takeuchi, *The Knowledge-Creating Company: How Japanese Companies Create the Dynamics of Innovation* (New York: Oxford University Press, 1995); K. Sveiby, *The New Organizational Wealth* (San Francisco, CA: Berret-Koehler, 1997).
2. M. Alavi and M. D. Leidner, 'Knowledge Management and Knowledge Management Systems: Conceptual Foundations and Research Issues', *MIS Quarterly*, 25 (2001), pp. 107–36.

3. I. Nonaka and V. Peltokorpi, 'Objectivity and Subjectivity in Knowledge Management: A Review of 20 Top Articles', *Knowledge and Process Management*, 13:2 (2006), pp. 73–82.

4. G. Schreyögg and D. Geiger, 'Kann implizites Wissen Wissen sein? Vorschläge zur Neuorientierung im Wissensmanegement', in B. Wyssusek (ed.), *Wissensmanagement komplex* (Berlin: Erich Schmidt Verlag, 2004), pp. 43–54.

5. For example, J. Stanley and T. Williamson, 'Knowing How', *Journal of Philosophy*, 98:8 (2001), pp. 411–44.

6. U. V. Riss, 'Knowledge, Action, and Context: Impact on Knowledge Management', in K.-D. Althoff et al. (eds), *Professional Knowledge Management, Lecture Notes in Artificial Intelligence* (Berlin: Springer, 2005), pp. 598–608, and references there.

7. I. Nonaka, R. Toyama and A. Nagata, 'A Firm as a Knowledge-Creating Entity: A New Perspective on the Theory of the Firm', *Industrial and Corporate Change*, 9:1 (2000), pp. 1–20.

8. I. Nonaka and R. Toyama, 'The Knowledge-Creating Theory Revisited: Knowledge Creation as a Synthesizing Process', *Knowledge Management Research and Practice*, 1 (2003), pp. 2–10.

9. A. Kern, *Quellen des Wissens* (Frankfurt: Suhrkamp, 2006).

10. P. Ruben, *Dialektik und Arbeit der Philosophie* (Cologne: Pahl-Rugenstein, 1978).

11. L. Wittgenstein, *Philosophical Investigations* (Oxford: Blackwell, 1953).

12. Ruben, *Dialektik und Arbeit der Philosophie*, pp. 52–98.

13. M. Williams, *Problems of Knowledge* (Oxford: Oxford University Press, 2001).

14. E. Gettier, 'Is Justified True Belief Knowledge?', *Analysis*, 23:6 (1963), pp. 121–3.

15. R. Keefe, *Theories of Vagueness* (Cambridge: Cambridge University Press, 2000).

16. P. Ruben, 'Der dialektische Widerspruch', in U. Hedtke, C. Warneke (eds.), *Peter Ruben – Philosophische Schriften*, online edn, at http://www.peter-ruben.de/frames/files/Grundlagen/Ruben%20-%20Der%20dialektische%20Widerspruch.pdf [accessed 7 March 2014].

17. G. W. F. Hegel, *Vorlesungen über die Geschichte der Philosophie I* (Frankfurt: Suhrkamp, 1971).

18. R. Hagengruber and U. V. Riss, 'Knowledge in Action', in G. Dodig-Crnkovic and S. Stuart (eds), *Computation, Information, Cognition: The Nexus and The Liminal* (Newcastle upon Tyne: Cambridge Scholars Publishing, 2007).

19. Concrete contradictions are also described as dialectical contradiction, as we find in the law of dialectical contradiction.

20. H. Pietschmann, *Phänomenologie der Naturwissenschaften* (Berlin: Springer, 1996).

21. B. Russell, *Principles of Mathematics* (Cambridge: Cambridge University Press, 1903).

22. K. Popper, *The Logic of Scientific Discovery* (London: Hutchinson, 1959).

23. K. Popper, 'What is Dialectic?', in K. Popper, *Conjecture and Refutation* (London: Routledge and Keagan, 1963), pp. 312–35.

24. The term 'subsidiary awareness' is taken from M. Polanyi.

25. A. S. Reber, 'Implicit Learning and Tacit Knowledge', *Journal of Experimental Psychology: General*, 118 (1989), pp. 219–35.

26. J. Habermas, 'Handlungen, Operationen, körperliche Bewegungen', in J. Habermas, *Vorstudien und Ergänzungen zur Theorie des kommunikativen Handelns* (Frankfurt: Suhrkamp, 1989), pp. 273–306.

27. For example, this contrasts with Habermas, who argued for a general footing of action on rules.

28. H. Collins and M. Kusch, *The Shape of Actions: What Humans and Machines Can Do* (Cambridge, MA: MIT Press, 1998); Habermas, 'Handlungen, Operationen, körperliche Bewegungen'.

29. K. Hawley, 'Success and Knowledge-How', *American Philosophical Quarterly*, 40 (2003), pp. 19–31.

30. J. Stanley, *Knowledge and Practical Interests* (Oxford: Oxford University Press, 2007).

31. G. Walsham, 'Knowledge Management Systems: Representation and Communication in Context', *Systems, Signs and Actions*, 1 (2005), pp. 6–18.

32. M. Polanyi, *Knowing and Being* (Chicago, IL: University of Chicago Press, 1969).

33. The general approach is described in U. V. Riss, 'Pattern-Based Task Management as Means of Organizational Knowledge Maturing', *International Journal of Knowledge-Based Organizations*, 1:1 (2011), pp. 20–41; action theory is described in L. Vygotsky, *Mind in Society: Development of Higher Psychological Processes* (Cambridge, MA, and London: Harvard University Press, 1978) and Y. Engeström, *Learning by Expanding: An Activity-Theoretical Approach to Developmental Research* (Helsinki: Orienta-Konsultit Oy, 1987).

34. U. V. Riss, O. Grebner and Y. Du, 'Task Journals as Means to Describe Temporal Task Aspects for Reuse in Task Patterns', in *Proceedings of the 9th European Conference on Knowledge Management, ECKM 2008* (Amsterdam: Elsevier, 2008), pp. 721–9.

35. E. Ong et al., 'Semantic Task Management Framework: Bridging Information and Work', in T. Pellegrini (ed.), *Networked Knowledge – Networked Media* (Berlin: Springer, 2009).

36. Riss, Grebner and Du, 'Task Journals as Means to Describe Temporal Task Aspects for Reuse in Task Patterns'.

37. O. Grebner, E. Ong and U. V. Riss, 'KASIMIR – Work Process Embedded Task Management Leveraging the Semantic Desktop', in M. Bichler (ed.), *Multikonferenz Wirtschaftsinformatik 2008* (Berlin: GITO-Verlag 2008), pp. 715–26.

38. F. Dudda, 'Gettier-Beispiele und eine Gebrauchsdefinition des Begriffs des propositionalen Wissens', *Facta Philosophica*, 9:1 (2007), pp. 161–76.

39. D. Davidson and H. F. Fulda, *Dialektik und Dialog* (Frankfurt: Suhrkamp, 1993).

14 Holzweißig and Krüger, 'Action-Directing Construction of Reality in Product Creation Using Social Software: Employing Philosophy to Solve Real-World Problems'

1. W. J. Ohms, *Management des Produktentstehungsprozesses* (Munich: Franz Vahlen, 2000).

2. R. G. Cooper, 'Third-Generation New Product Processes', *Journal of Product Innovation Management*, 11:1 (1994), pp. 3–14; R. G. Cooper, *Winning at New Products: Accelerating the Process from Idea to Launch*, 3rd edn (Cambridge: Perseus, 2001).

3. R. Hagenruber and U. V. Riss, 'Knowledge in Action', in G. Dodig-Crnkovic and S. Stuart (eds), *Computation, Information, Cognition: The Nexus and the Liminal* (Newcastle upon Tyne: Cambridge Scholars Press, 2007), pp. 134–47.

4. J. R. Searle, *The Construction of Social Reality* (New York: The Free Press, 1995); J. R. Searle, 'Social Ontology: Some Basic Principles', *Anthropological Theory*, 6:1 (2006), pp. 12–29.

5. Searle's concept of collective intentionality shows strong parallels to Tuomela's works on 'we-attitudes' (mutual belief, joint intentions, etc.) Cf. R. Tuomela, *A Theory of Social Action* (Dordrecht: Reidel, 1984).

6.   Cf. P. Watzlawick (ed.), *Die erfundene Wirklichkeit: Wie wissen wir, was wir zu wis-sen glauben?*, 17th edn (Munich: Piper, 2004); E. von Glasersfeld, *The Construction of Knowledge* (Salinas: Intersystems Publications, 1988).

7.   D. Dougherty, 'Interpretative Barriers to Successful Product Innovation in Large Firms', *Organization Science*, 3:2 (1992), pp. 179–202.

8.   B. Wyssusek, *Methodologische Aspekte der Organisationsmodellierung* (Saarbrücken: Ver-lag Dr. Müller, 2008).

9.   Works in informatics which employ an interpretative approach – such as the one by Kremberg, Wyssusek and Schwartz involving software development and reality construc-tion – are more an exception than the general rule. Cf. B. Kremberg, B. Wyssusek and M. Schwartz, 'Soziopragmatischer Konstruktivismus und Softwaretechnik: Eine kritische Analyse des Informationsbegriffes als Grundlage für das Verständnis von Informations-systemen', in K. Bauknecht, W. Brauer and T. Mück (eds), *Informatik 2001: Wirtschaft und Wissenschaft in der Network Economy – Visionen und Wirklichkeit*, Tagungsband der GI/OCG-Jahrestagung (Vienna: Österreichische Computer Gesellschaft, 2001), pp. 748–54.

10. Cf. S. Wolf, *Wissenschaftstheoretische und fachmethodische Grundlagen der Konstruktion von generischen Referenzmodellen betrieblicher Systeme* (Aachen: Shaker Verlag, 2001).

11. Cf. Wyssusek, *Methodologische Aspekte der Organisationsmodellierung*.

12. Kremberg, Wyssusek and Schwartz, 'Soziopragmatischer Konstruktivismus und Soft-waretechnik'.

13. W. K. Köck, 'Kognition – Semantik – Kommunikation', in S. J. Schmidt (ed.), *Der Dis-kurs des Radikalen Konstruktivismus* (Frankfurt: Suhrkamp, 1987), pp. 340–73.

14. Kremberg, Wyssusek and Schwartz, 'Soziopragmatischer Konstruktivismus und Soft-waretechnik'.

15. 'Self-explanatory', in our view, is a philosophical faux-pas term, as is noticeable from our explications in this essay. Someone can only speak about something in a sense-full man-ner, if he or she already possesses precognitions on that matter. Hence, our notion is that something cannot be understood as self-explanatory unless the reader has similar precognitions and presuppositions to those of the author.

16. Dougherty, 'Interpretative Barriers to Successful Product Innovation in Large Firms'.

17. Here, we refer to 'one language' in terms of having compatible interpretative schemes that yield similar attributions of meaning while employing the same natural language.

18. According to Boster, cognitive anthropology is defined as the 'study of the *content* of thought (or knowledge) in *communities* of individuals observed in *natural* settings. Cognitive anthropologists seek to understand how culture happens, to explore how col-lective understandings of the world emerge in social groups, and to discover the pattern of cross-cultural similarities and differences in culture and cognition.' J. S. Boster, 'Cat-egories and Cognitive Anthropology', in H. Cohen and C. Lefebvre (eds), *Handbook of Categorization in Cognitive Science*, 1st edn (Amsterdam: Elsevier, 2005), pp. 91–118, on p. 93.

19. Cf. ibid; J. S. Boster, 'Emotion Categories across Languages', in Cohen and Lefebvre (eds), *Handbook of Categorization in Cognitive Science*, 1st edn, pp. 187–222.

20. M. W. Eysenck and M. T. Keane, *Cognitive Psychology: A Student's Handbook*, 5th edn (Hove: Psychology Press, 2005).

21. H. von Foerster, 'Das Konstruieren einer Wirklichkeit', in Watzlawick (ed.), *Die erfundene Wirklichkeit*, 17th edn, pp. 39–66.

22. Cf. Searle, *The Construction of Social Reality*; C. K. Eoyang, 'Symbolic Transformation of Belief Systems', in L. R. Pondy et al. (eds), *Organizational Symbolism* (Greenwich: Jai Press, 1983), pp. 109–21.

23. P. L. Berger and T. Luckmann, *Die gesellschaftliche Konstruktion der Wirklichkeit*, 21st edn (Frankfurt: Fischer, 2007).

24. This point of view shows strong parallels with the theory of symbolic interactionism (TSI), in which action processes are thought to employ significant symbols that, through reciprocal employment by all actors, create a common orientation; cf. H. J. Helle, *Theorie der symbolischen Interaktion*, 3rd edn (Wiesbaden: Westdeutscher Verlag, 2001). In this way, habitualized patterns of (inter)action give rise to social institutions – social action is seen as the sense-constituting element.

25. D. Krause, *Luhmann-Lexikon: Eine Einführung in das Gesamtwerk von Niklas Luhmann* (Stuttgart: Lucius und Lucius, 2001).

26. Cf. P. N. Johnson-Laird, *Mental Models: Towards a Cognitive Science of Language, Inference, and Consciousness, Cognitive Science Series Vol. 6* (Cambridge, MA: Harvard University Press, 1983); K. Langfield-Smith, 'Exploring the Need for a Shared Cognitive Map', *Journal of Management Studies*, 29 (1992), pp. 349–68.

27. Krause, *Luhmann-Lexikon*.

28. C. Hardy, T. B. Lawrence and D. Grand, 'Discourse and Collaboration: The Role of Conversations and Collective Identity', *Academy of Management Review*, 30 (2005), pp. 58–77, on p. 60.

29. N. Fairclough, *Discourse and Social Change* (Cambridge: Polity Press, 1992), p. 60.

30. Hardy, Lawrence and Grand, 'Discourse and Collaboration: The Role of Conversations and Collective Identity', *Academy of Management Review*, 30 (2005), pp. 58–77.

31. Ibid.

32. A. Komus and F. Wauch, *Wikimanagement: Was Unternehmen von Social Software und Web 2.0 lernen können* (Munich: Oldenbourg, 2008).

33. C. Pentzold, 'Machtvolle Wahrheiten. Diskursive Wissensgenerierung in Wikipedia aus Foucault'scher Perspektive', in C. Stegbauer et al. (eds), *Wikis: Diskurse, Theorien und Anwendungen*, Kommunikation@Gesellschaft, 8 (2007), Online-Publikation, at http://www.soz.uni-frankfurt.de/K.G/B4_2007_Pentzold.pdf [accessed 7 March 2014].

34. R. Reagans and B. McEvily, 'Network Structure and Knowledge Transfer: The Effects of Cohesion and Range', *Administrative Science Quarterly*, 48 (2003), pp. 240–67.

35. We thank Barry Smith for helpful advice on this issue.

15 Pross, 'An Action-Theory-Based Treatment of Temporal Individuals'

1. In the sense of Z. Vendler, 'Verbs and Times', *Philosophical Review*, 66:2 (1957), pp. 143–60; D. Davidson, 'The Logical Form of Action Sentences', in N. Rescher (ed.), *The Logic and Decision of Action* (Pittsburgh, PA: The University of Pittsburgh Press, 1967), pp. 81–95.

2. See D. Dennett, *The Intentional Stance* (Cambridge, MA: MIT Press, 1989); F. Dretske, *Explaining Behavior: Reasons in a World of Causes* (Cambridge, MA: MIT Press, 1988); and D. Hartmann and P. Janich, *Methodischer Kulturalismus* (Frankfurt: Suhrkamp, 1991).

3.   H. Kamp, J. van Genabith and U. Reyle, 'Discourse Representation Theory', in D. Gabbay and F. Guenthner (eds), *Handbook of Philosophical Logic* (Dordrecht: Kluwer, 2007).
4.   Davidson, 'The Logical Form of Action Sentences'.
5.   H. Kamp, 'Intentions, Plans and their Execution: Turning Objects of Thought into Entities of the External World', unpublished manuscript, IMS University of Stuttgart, 2007, at http://www.uni-stuttgart.de/linguistik/sfb732/files/Kamp_Intentions.pdf [accessed 7 March 2014].
6.   For reasons of space and time, I cannot discuss the interactions between tense, grammatical aspects and lexical aspects in detail here.
7.   Vendler, 'Verbs and Times'.
8.   M. Moens and M. Steedman, 'Temporal Ontology and Temporal Reference', *Computational Linguistics*, 14 (1988), pp. 15–28.
9.   There are, of course, more sensible theories about creation verbs with respect to accomplishments in DRT, for example in the unpublished manuscript by H. Kamp and A. Bende-Farkas, 'Verbs of Creation' (IMS University of Stuttgart, 2005). Thus the representation pictured here certainly does not give a fair depiction of the current state of DRT. Nevertheless, the current argumentation applies to these refinements too.
10. D. R. Dowty, *Word Meaning und Montague Grammar* (New York: Springer, 1979).
11. M. van Lambalgen and F. Hamm, *The Proper Treatment of Events* (Oxford: Blackwell, 2004).
12. The use of pre-states and post-states has to face a lot more problems than I can discuss here, i.e. the amount of information contained in these states as well as their temporal extent.
13. Given the problems of meaning postulates as spelled out above, I do not consider them as providing instrumental information.
14. Moens and Steedman, 'Temporal Ontology and Temporal Reference'.
15. J. R. Searle, *Speech Acts: An Essay in the Philosophy of Language* (Cambridge: Cambridge University Press, 1969).
16. J. M. Zacks and K. M. Swallow, 'Event Segmentation', *Current Directions in Psychological Science*, 16:2 (2007), pp. 80–4; J. M. Zacks and B. Tversky, 'Event Structure in Perception and Conception', *Psychological Bulletin*, 127 (2001), pp. 3–21.
17. This seems to be the right point at which to make a note about my jargon: the term 'temporal variation' as I understand it refers to an uninterpreted, unsegmented sequence of action from which no temporal entities such as events have been extracted. Segmentation then establishes temporal profiles, i.e. structures on the temporal varation. Finally, a temporal entity represents a temporal profile (or higher-order constructions of temporal profiles).
18. D. Davidson, 'Actions, Reasons and Causes', *Journal of Philosophy*, 60 (1963), pp. 695–700.
19. E. Anscombe, *Intention* (Oxford: Basil Blackwell, 1957).
20. Dennett, *The Intentional Stance*.
21. Dretske, *Explaining Behavior*.
22. Hartmann and Janich, *Methodischer Kulturalismus*.
23. M. E. Bratman, *Intention, Plans, and Practical Reason* (Cambridge, MA: Harvard University Press, 1987).
24. E. A. Emerson, 'Temporal and Modal Mogic', in J. van Leeuwen (ed.), *Handbook of Theoretical Computer Science* (Amsterdam: North-Holland Publishing Company, 1990), p. 1072.

25. The notion of 'atomic' requires a note on the granularity of atomic actions. In this essay, the atomic actions have a quite high granularity that should be lowered for reasonable results.

26. Usually one would suppose that plans are not just sequences of action, but that they involve decisions for one or the other option of action based on the epistemic states of the agent and the current states of affairs. For reasons of space and simplicity, I assume that plans are sequences of action whereas a not so simple-minded approach would use programme-like planning structures. For example, see M. P. Singh, *Multiagent Systems: A Theoretical Framework for Intentions, Know-How and Communications* (New York: Springer, 1994); and M. D'Inverno et al., 'The dMARS Architecture: A Specification of the Distributed Multi-Agent Reasoning System', *Autonomous Agents and Multi-Agent Systems*, 9 (2004), pp. 5–53.

27. As intentions are inherently tied to actions, probably the most natural solution to intentions should consult an operational semantics defined in terms of algorithmic specifications: how intentions result from desires by means of choice and commitment. For ease of exposition, I adopt a simplistic solution that treats desires and intentions by means of assignment functions.

28. T. Pross, 'Grounded Discourse Representation Theory: Toward a Semanticspragmatics Interface for Human–Machine Collaboration' (PhD dissertation, IMS University of Stuttgart, 2009).

16 Jansen, 'Four Rules for Classifying Social Entities'

1. See http://www.ifomis.org/bfo [accessed 7 March 2014].

2. M. Gilbert, *On Social Facts* (London: Routledge, 1989).

3. R. Tuomela, *The Importance of Us: A Philosophical Study of Basic Social Notions* (Stanford, CA: Stanford University Press, 1995).

4. J. R. Searle, *The Construction of Social Reality* (New York: The Free Press, 1995).

5. B. Smith, 'From Speech Acts to Social Reality', in B. Smith (ed.), *John Searle* (Cambridge: Cambridge University Press, 2003), pp. 1–33; B. Smith and J. R. Searle, 'The Construction of Social Reality: An Exchange', *American Journal of Economics*, 62 (2003), pp. 285–309, also published in L. Moss and D. Koepsell (eds), *Searle on the Institutions of Social Reality* (Oxford: Blackwell, 2003).

6. The NCIT can be freely accessed at http://bioportal.nci.nih.gov/ [accessed 7 March 2014]. All references in this paper are to version 08.06d.

7. S. Coronado et al., 'NCI Thesaurus: Using Science-Based Terminology to Integrate Cancer Research Results', in M. Fieschi et al. (eds), *Proceedings of MedInfo 2004* (Amsterdam: IOS Press, 2004), pp. 33–7; G. Fragoso, 'Overview and Utilization of the NCI Thesaurus', *Comperative and Functional Genomics*, 5 (2004), pp. 648–54; J. Golbeck, 'The National Cancer Institute's Thesaurus and Ontology', *Journal of Web Semantics*, 1 (2004), pp. 75–80.

8. NCICB (NCI Center for Bioinformatics), 'The NCICB User Applications Manual 2004', at https://ncicbsupport.nci.nih.gov/sw/content/NCICBAppManual.pdf [accessed 27 September 2013], p. 14.

9. For other criticisms, see L. Jansen, 'Classifications', in K. Munn and B. Smith (eds), *Applied Ontology: An Introduction* (Frankfurt: Ontos, 2008), pp. 159–72. The German version is available in L. Jansen and B. Smith (eds), *Biomedizinische Ontologie. Wissen strukturieren für den Informatik-Einsatz* (Zurich: vdf, 2008), pp. 67–83.

10. Fragoso, 'Overview and Utilization of the NCI Thesaurus'.
11. G. A. Colditz, 'Introduction', *Encyclopedia of Cancer and Society*, 3:1 (2007), pp. vii–ix.
12. Coronado et al., 'NCI Thesaurus'.
13. N. Guarino and C. Welty, 'An Overview of OntoClean', in S. Staab and R. Studer (eds), *The Handbook on Ontologies* (Berlin: Springer, 2004), pp. 151–72.
14. B. Smith et al., 'Relations in Biomedical Ontologies', *Genome Biology*, 6:5 (2005), at http://genomebiology.com/content/pdf/gb-2005-6-5-r46.pdf [accessed 27 September 2013].
15. U. Schwarz and B. Smith, 'Ontological Relations', in Munn and Smith (eds), *Applied Ontology*, pp. 155–72.
16. See A. Galton, *The Logic of Aspect* (Oxford: Clarendon Press, 1984), pp. 153–6.
17. J. P. Pickett et al. (eds), *Fund: The American Heritage Dictionary of the English Language* (Boston, MA: Houghton Mifflin, 2000).
18. For more on this, see O. Bodenreider, B. Smith and A. Burgun, 'The Ontology–Epistemology Divide: A Case Study in Medical Terminology', in A. Varzi and L. Vieu (eds), *Formal Ontology in Information Systems: Proceedings of the International Conference on Formal Ontology and Information Systems* (Amsterdam: IOS Press, 2004), pp. 185–95.
19. L. Jansen, 'Categories: The Top-Level-Ontology', in Munn and Smith (eds), *Applied Ontology*, pp. 173–96. The German version is in Jansen and Smith (eds), *Biomedizinische Ontologie*, pp. 85–112; B. Smith, 'Aristoteles 2002', in T. Buchheim, H. Flashar and R. A. H. King (eds), *Kann man heute noch etwas anfangen mit Aristoteles?* (Hamburg: Meiner, 2003), pp. 3–38; Smith, 'Relations in Biomedical Ontologies'.
20. Jansen, 'Categories: The Top-Level-Ontology'; A. D. Spear, 'Ontology for the Twenty-First Century: An Introduction with Recommendations', Saarbrücken, 2006, at http://www.ifomis.org/bfo/documents/manual.pdf [accessed 27 September 2013].
21. L. Jansen, 'Institutionen und die kategoriale Ontologie', in G. Schönrich (ed.), *Institutionen und ihre Ontologie* (Frankfurt and Lancaster: Ontos, 2005), pp. 45–57.
22. T. Kobusch, *Die Entdeckung der Person. Metaphysik der Freiheit und modernes Menschenbild* (Darmstadt: Wissenschaftliche Buchgesellschaft, 2007).
23. Jansen, 'Institutionen und die kategoriale Ontologie'.
24. Searle, *The Construction of Social Reality*.
25. Jansen, 'Institutionen und die kategoriale Ontologie'.
26. L. Jansen, 'Unity and Constitution of Social Entities', in L. Honnefelder and E. Runggaldier (eds), *Unity and Time in Metaphysics* (Berlin: de Gruyter, 2009), pp. 15–45.
27. L. Vizenor, 'Actions in Health Care Organizations: An Ontological Analysis', in M. Fieschi (eds), *Proceedings of MedInfo 2004* (Amsterdam: IOS, 2004), pp. 1403–8.
28. B. Bolzano, *Grundlagen der Logik* (Hamburg: Meiner, 1978), pp. 93–7.
29. Spear, *Ontology for the Twenty-First Century*, pp. 55–6.
30. R. Mizoguchi, 'The Model of Roles within an Ontology Development Tool: Hozo', *Applied Ontology*, 2 (2007), pp. 159–79.
31. F. Loebe, 'Abstract vs. Social Roles: Towards a General Theoretical Account of Roles', *Applied Ontology*, 2 (2007), pp. 127–58.
32. Spear, *Ontology for the Twenty First Century*, p. 43.
33. Schwarz and Smith, 'Ontological Relations'; Smith, 'Relations in Biomedical Ontologies'.
34. W. Ceusters, B. Smith and L. Goldberg, 'A Terminological Analysis and Ontological Analysis of the NCI Thesaurus', *Methods of Information in Medicine*, 44 (2005), pp. 498–507.

35. A. Quinton, 'Social Objects', *Proceedings of the Aristotelian Society*, 76 (1975–6), pp. 1–27.

36. D.-H. Ruben, *The Metaphysics of the Social World* (London: Routledge & Kegan Paul, 1985), ch. 2.

37. B. Smith and P. Grenon, 'The Cornucopia of Formal-Ontological Relations', *Dialectica*, 58 (2004), pp. 279–96; Smith, 'Relations in Biomedical Ontologies'.

17 Dodig-Crnkovic, 'Info-Computationalism and Philosophical Aspects of Research in Information Sciences'

1. T. Stonier, *The Wealth of Information* (London: Thames and Methuen, 1993).

2. G. J. Chaitin, 'Epistemology as Information Theory: Alan Turing Lecture given at E-CAP 2005', in G. Dodig-Crnkovic and S. Stuart (eds), *Computation, Information, Cognition: The Nexus and The Liminal* (Newcastle upon Tyne: Cambridge Scholars Publishing, 2007), pp. 2–17.

3. E. Pivčević, *The Reason Why: A Theory of Philosophical Explanation* (Hrvatski Leskovac: KruZak, 2007).

4. A. Betti and W. R. de Jong (eds), 'The Classical Model of Science I: A Millennia-Old Model of Scientific Rationality', *Synthese*, 174 (2010), pp. 185–203.

5. Ibid.

6. Chaitin, 'Epistemology as Information Theory'.

7. S. Lloyd, *Programming the Universe: A Quantum Computer Scientist Takes on the Cosmos* (New York: Alfred A. Knopf, 2006). See more about the question of digital/analogue universe in G. Dodig-Crnkovic, 'Investigations into Information Semantics and Ethics of Computing' (PhD dissertation, Mälardalen University, 2006).

8. G. Dodig-Crnkovic, 'Epistemology Naturalized: The Info-Computationalist Approach', *APA Newsletter on Philosophy and Computers*, 6:2 (2007), pp. 9–14.

9. G. Dodig-Crnkovic, 'Shifting the Paradigm of the Philosophy of Science: The Philosophy of Information and a New Renaissance', *Minds and Machines*, 13:4 (2003), pp. 521–36.

10. See for example Chaitin, 'Epistemology as Information Theory'; L. Floridi, 'Open Problems in the Philosophy of Information', *Metaphilosophy*, 35:4 (2004), pp. 554–82; L. Floridi (ed.), *The Blackwell Guide to the Philosophy of Computing and Information* (Oxford: Blackwell, 2004); Lloyd, *Programming the Universe*; K. Mainzer, *Thinking in Complexity* (Berlin: Springer, 2004); K. Mainzer, *Computerphilosophie zur Einführung* (Hamburg: Junius, 2003); S. Wolfram, 'A New Kind of Science', at http://www.wolfram-science.com/nksonline/toc.html [accessed 23 September 2013]; K. Zuse, 'Rechnender Raum', *Elektronische Datenverarbeitung* (1967), pp. 336–44.

11. Dodig-Crnkovic, 'Investigations into Information Semantics and Ethics of Computing'; G. Dodig-Crnkovic and S. Stuart (eds), *Computation, Information, Cognition: The Nexus and The Liminal* (Newcastle upon Tyne: Cambridge Scholars Publishing, 2007).

12. Lloyd, *Programming the Universe*.

13. Mainzer, *Thinking in Complexity*.

14. See for example J. Smith and C. Jenks, *Qualitative Complexity Ecology, Cognitive Processes and the Re-Emergence of Structures in Post-Humanist Social Theory* (London: Routledge, 2006).

15. This can also be found in Mainzer, *Thinking in Complexity*.

16. Alan Turing was one of the notable exemptions to the rule. Others are Weizenbaum, Winograd and Flores. It should also be mentioned that computing always had strong bonds with logic, and that AI in particular always had recognized philosophical aspects.

17. Dodig-Crnkovic, 'Shifting the Paradigm of the Philosophy of Science'.

18. See for example Floridi's lecture in the Swedish National PI course on the development of philosophy: PI, 'Swedish National Course', at http://www.idt.mdh.se/~gdc/PI_04/index.html [accessed 23 September 2013].

19. L. Wolpert, *The Unnatural Nature of Science* (London: Faber 1992).

20. J. Ladyman and D. Ross, *Every Thing Must Go: Metaphysics Naturalized* (Oxford: Oxford University Press, 2007).

21. J. Moor, 'What is Computer Ethics?', *Metaphilosophy*, 16:4 (1985), pp. 266–75.

18 Müller, 'Pancomputationalism: Theory or Metaphor?'

1. D. Adams, *The Ultimate Hitchhiker's Guide to the Galaxy* (New York: Del Rey, 1979).

2. I believe the term pancomputationalism was introduced in G. Dodig-Crnkovic, 'Epistemology Naturalized: The Info-Computationalist Approach', *APA Newsletter on Philosophy and Computers*, 6:2 (2007), pp. 9–14.

3. D. J. Chalmers, 'A Computational Foundation for the Study of Cognition', at http://consc.net/papers/computation.html [accessed 23 September 2013]; D. J. Chalmers, 'On Implementing a Computation', *Minds and Machines*, 4 (1994), pp. 391–402; D. J. Chalmers, 'Does a Rock Implement Every Finite-state Automaton', *Synthese*, 108 (1996), pp. 309–30.

4. J. R. Searle, 'Minds, Brains and Programs', *Behavioral and Brain Sciences*, 3 (1980), pp. 417–57; J. R. Searle, 'Is the Brain a Digital Computer?', *Proceedings and Addresses of the American Philosophical Association*, 64 (1990), pp. 21–37; J. R. Searle, *The Rediscovery of Mind* (Cambridge, MA: MIT Press, 1992).

5. O. Shagrir, 'More on Global Supervenience', *Philosophy and Phenomenological Research*, 59:3 (1999), pp. 691–701.

6. For the beginnings of an argument, see my previous considerations: V. C. Müller, 'Representation in Digital Systems', in A. Briggle, K. Waelbers and P. Brey (eds), *Current Issues in Computing and Philosophy* (Amsterdam: IOS Press 2008), pp. 116–21.

7. Dodig-Crnkovic, 'Epistemology Naturalized'.

8. G. Chatin, 'Epistemology as Information Theory: From Leibniz to Ω', in G. Dodig-Crnkovic and S. Stuart (eds), *Computation, Information, Cognition: The Nexus and the Luminal* (Newcastle upon Tyne: Cambridge Scholars Publishing 2007), pp. 27–51.

9. Ibid.

10. E. Fredkin, 'An Introduction to Digital Philosophy', *International Journal of Theoretical Physics*, 42:2 (2003), pp. 189–247.

11. S. Wolfram, 'A New Kind of Science', at http://www.wolframscience.com/nksonline/toc.html [accessed 23 September 2013].

12. K. Kelly, 'God is the Machine', at http://www.wired.com/wired/archive/10.12/holytech.html [accessed 23 September 2013].

13. M. Miłkowski, 'Is Computationalism Trivial?', in Dodig-Crnkovic and Stuart (eds), *Computation, Information, Cognition*, pp. 236–46.

14. J. Kim, 'Concepts of Supervenience', *Philosophy and Phenomenological Research*, 45:2 (1984), pp. 153–76.

15. Shagrir, 'More on Global Supervenience'.

16. L. Floridi, 'The Method of Levels of Abstraction', *Minds and Machines*, 18:3 (2008), pp. 303–29; Müller, 'Representation in Digital Systems'.

17. Searle, 'Minds, Brains and Programs'.

18. D. Deutsch, 'It from Qubit', in J. D. Barrow, P. C. D. Davies and C. L. Harper (eds), *Science and Ultimate Reality: Quantum Theory, Cosmology, and Complexity* (Cambridge: Cambridge University Press, 2004), pp. 90–102.

19. H. Putnam, 'Models and Reality', in H. Putnam, *Realism and Reason: Philosophical Papers Volume 3* (Cambridge: Cambridge University Press, 1980), pp. 1–25.

19 Dane, 'The Importance of the Sources of Professional Obligations'

1. B. Russel, *The Problems of Philosophy* (Oxford: Oxford University Press, 1912).

2. J. Chaffee, *The Philosopher's Way: Thinking Critically About Profound Ideas* (Upper Saddle River: Pearson Prentice Hall, 2005).

3. See F. C. Dane, *Research Methods* (Pacific Grove: Brooks/Cole Publishing Company, 1990); F. C. Dane, 'Ethical Issues in Statistical Analyses: An Argument for Collective Moral Responsibility', *C-Stat News*, 2:3 (2006), pp. 1, 3, 4; F. C. Dane, 'Ethics of Stem-Cell Research: A Framework for Ethical Dialogue Regarding Sources of Conflict', in M. N. Bahvani (ed.), *Stem Cell Research: Law and Ethics* (Hyderabad: Amicus Books, 2009), pp. 62–70; F. C. Dane, 'Evaluation Research', in S. Zhongxin and L. Zhang (eds), *Feminist Research Methods* (Shanghai: Fudan University Press, 2008), pp. 188–219; F. C. Dane and D. C. Parish, 'Ethical Issues in Registry Research: In-Hospital Resuscitation as a Case Study', *Journal of Empirical Research on Human Research Ethics*, 1:4 (2006), pp. 69–75.

4. See Chaffee, *The Philosopher's Way*; O. A. Johnson (ed.), *Ethics: Selections from Classical and Contemporary Writers* (New York: Holt, Rinehart and Winston, 1989); P. Singer, *Writings on an Ethical Life* (New York: HarperCollins, 2000); J. Ladd, 'Collective and Individual Moral Responsibility in Engineering: Some Questions', *Ethical Issues in Engineering. IEEE Technology and Society Magazine*, 1:2 (1992), pp. 3–10.

5. F. C. Dane, 'Academic Integrity', *SVSU Literacy Link*, 9:1 (2004), pp. 3–4; Dane, 'Ethical Issues in Statistical Analyses', pp. 1, 3, 4; F. C. Dane, 'Restoring Trust in Government and Business: Principles over Rules', in K. Ballstadts (ed.), *Oxford Round Table, Ethical Sentiments: The Restoration of Trust in Government and Business*, forthcoming.

6. Center for the Study of Ethics in the Professions at IIT, *Codes of Ethics Online* (Chicago, IL: Illinois Institute of Technology, 15 March 2008); Dane, 'Restoring Trust in Government and Business: Principles over Rules'.

7. S. H. Unger, 'Codes of Engineering Ethics: Ethical Issues in Engineering', in D. G. Johnson (ed.), *Controlling Technology: Ethics and the Responsible Engineer* (Upper Saddle River: Prentice Hall, 1991), pp. 105–29.

8. Ladd, 'Collective and Individual Moral Responsibility in Engineering'.

9. 'Software Engineering Code of Ethics and Professional Practice (Version 5.2)', at http://http://www.acm.org/about/se-code/#full [accessed 20 September 2013].

10. Ibid.

11. D. G. Johnson, *Computer Ethics* (Upper Saddle River: Prentice Hall, 2001). D. G. Johnson, 'The Social and Professional Responsiblity of Engineers: Ethical Issues in Engineering', *Annals of New York Academy of Sciences*, 577 (1989), pp. 106–14.

12. J. Bentham, *An Introduction to the Principles of Morals and Legislation* (1907; Oxford: Clarendon Press, 2002).

13. The second aspiration within the Code, to act in the best interests of clients and employers 'consistent with the public interest', can be analysed equally well from an utilitarian viewpoint, but clearly places the public interest as primary over those of one's clients and employers. See 'Software Engineering Code of Ethics and Professional Practice (Version 5.2)'.

14. Bentham, *An Introduction to the Principles of Morals and Legislation*.

15. Dane, 'Academic Integrity'.

16. Chaffee, *The Philosopher's Way*.

17. Aristotle, *Nicomachean Ethics*, The Internet Classics Archive 2000, at http://classics.mit.edu/Aristotle/nicomachaen.html [accessed 20 September 2013].

18. Ibid.

19. J. Badaracco, *Defining Moments: When Managers Must Choose between Right and Right* (Cambridge, MA: Harvard Business Press, 1997); K. Kristjánsson, 'Emotional Intelligence in the Classroom? An Aristotelian Critique', *Educational Theory*, 56:1 (2006), pp. 39–56.

20. Aristotle, *Nicomachean Ethics*.

21. See R. L. Walker and P. J. Ivanhoe (eds), *Working Virtue* (Oxford: Oxford University Press, 2007).

22. I. Kant, *Critique of Practical Reason* (Cambridge: Cambridge University Press, 1997); I. Kant, *Groundwork of the Metaphysics of Morals* (Radford: Wilder Publications, 2008).

23. Kant, *Groundwork of the Metaphysics of Morals*, p. 39.

24. Ibid., p. 59.

25. J. Ladd, 'The Ethics of Participation', in J. R. Pennock and J. W. Chapman (eds), *Nomos Xvi: Participation in Politics* (New York: Atherton-Lieber, 1975), pp. 98–125; Ladd, 'Collective and Individual Moral Responsiblity in Engineering'; J. Ladd, 'The Quest for a Code of Professional Ethics: An Intellectual and Moral Confusion', in D. G. Johnson (ed.), *Ethical Issues in Engineering* (Englewood Cliffs, NJ: Prentice Hall, 1991), pp. 130–6.

26. C. M. Koorsgard, 'The Myth of Egoism', in P. Baumann (ed.), *Practical Conflicts: New Philosophical Essays* (Cambridge: Cambridge University Press, 2004), pp. 57–91.

27. Ladd, 'The Ethics of Participation'.

28. For example S. Stark, 'A Change of Heart: Moral Emotions, Transformation, and Moral Virtue', *Journal of Moral Philosophy: An International Journal of Moral, Political and Legal Philosophy*, 1:1 (2004), pp. 31–50.

29. R. Pozzo, 'Kant on the Five Intellectual Virtues', in R. Pozzo (ed.), *Studies in Philosophy and the History of Philosophy* (Washington, DC: Catholic University of America Press, 2004), pp. 173–92.

30. O. Höffe, 'Ethik ohne und mit Metaphysik: Zum Beispiel Aristoteles und Kant', *Zeitschrift für philosophische Forschung*, 61:4 (2007), pp. 405–22; J. B. Murphy, 'Practical Reason and Moral Psychology in Aristotle and Kant', *Social Philosophy and Policy*, 18:2 (2001), pp. 257–99.

31. Google Programming Contest, Mountain View 2002, at http://mooglemb.com/threads/googlecache/64.233.187.104/programming-contest/default.htm [accessed 20 September 2013].

32. J. Rawls, *A Theory of Justice* (Cambridge, MA: Harvard University Press, 1971); J. Rawls, *Political Liberalism* (New York: Columbia University Press, 1996); J. Rawls, *Justice as Fairness: A Restatement* (Cambridge, MA: Harvard University Press, 2001).

33. Rawls, *Justice as Fairness*.

34. Ibid.
35. M. C. Nussbaum, 'Cultivating Humanity', *Liberal Education* (1998), pp. 38–45. M. C. Nussbaum, *Women and Human Development* (Cambridge: Cambridge University Press, 2000).

INDEX